The Affective Negotiation of Slum Tourism

Each year, approximately a million tourists visit slum areas on guided tours as a part of their holiday to Asia, Africa or Latin America. This book analyses the cultural encounters that take place between slum tourists and former street children, who work as tour guides for a local NGO in Delhi, India.

Slum tours are typically framed as both *tourist performances*, bought as commodities for a price on the market, and as *appeals for aid* that tourists encounter within an altruistic discourse of charity. This book enriches the tourism debate by interpreting tourist performances as affective economies, identifying tour guides as emotional labourers and raising questions on the long-term impacts of economically unbalanced encounters with representatives of the Global North, including the researcher.

This book studies the 'feeling rules' governing a slum tour and how they shape interactions. When do guides permit tourists to exoticise the slum and feel a thrilling sense of disgust towards the effects of abject poverty, and when do they instead guide them towards a sense of solidarity with the slum's inhabitants? What happens if the tourists rebel and transgress the boundaries delimiting the space of comfortable affective negotiation constituted by the guides? This book will be essential reading for undergraduates, postgraduates and researchers working within the fields of human geography, slum tourism research, subaltern studies and development studies.

Tore Holst is External Lecturer at Cultural Encounters, Roskilde University, where he teaches mobility, migration, postcolonial literature, epistemology and the correlation between modernity and colonialism. He obtained his PhD from Roskilde University in 2016, with a thesis which this book is based on. He has also published postcolonial literature and focused on how the colonial relation between the Danish state and Greenland becomes visible when climate narratives are enacted and disseminated via the media.

Routledge Advances in Tourism and Anthropology

Series Edited by Dr Catherine Palmer and Dr Jo-Anne Lester
Dr Catherine Palmer (University of Brighton, UK) C.A.Palmer@brighton.ac.uk
Dr Jo-Anne Lester (University of Brighton, UK) J.Lester@brighton.ac.uk

To discuss any ideas for the series please contact Faye Leerink, Commissioning Editor: faye.leerink@tandf.co.uk or the Series Editors.

This series draws inspiration from anthropology's overarching aim to explore and better understand the human condition in all its fascinating diversity. It seeks to expand the intellectual landscape of anthropology and tourism in relation to how we understand the experience of being human, providing critical inquiry into the spaces, places, and lives in which tourism unfolds. Contributions to the series will consider how such spaces are embodied, imagined, constructed, experienced, memorialized and contested. The series provides a forum for cutting-edge research and innovative thinking from tourism, anthropology, and related disciplines such as philosophy, history, sociology, geography, cultural studies, architecture, the arts, and feminist studies.

The Affective Negotiation of Slum Tourism
City Walks in Delhi
Tore Holst

Tourism and Ethnodevelopment
Inclusion, Empowerment and Self Determination
Edited by Ismar Borges de Lima and Victor King

Everyday Practices of Tourism Mobilities
Packing a Bag
Kaya Barry

Tourism and Indigenous Heritage in Latin America
Casper Jacobsen

The Affective Negotiation of Slum Tourism
City Walks in Delhi

Tore Holst

LONDON AND NEW YORK

First published 2018
by Routledge
2 Park Square, Milton Park, Abingdon, Oxon OX14 4RN

and by Routledge
711 Third Avenue, New York, NY 10017

Routledge is an imprint of the Taylor & Francis Group, an informa business

© 2018 Tore Holst

The right Tore Holst to be identified as author of this work has been asserted by him in accordance with sections 77 and 78 of the Copyright, Designs and Patents Act 1988.

All rights reserved. No part of this book may be reprinted or reproduced or utilised in any form or by any electronic, mechanical, or other means, now known or hereafter invented, including photocopying and recording, or in any information storage or retrieval system, without permission in writing from the publishers.

Trademark notice: Product or corporate names may be trademarks or registered trademarks, and are used only for identification and explanation without intent to infringe.

British Library Cataloguing-in-Publication Data
A catalogue record for this book is available from the British Library

Library of Congress Cataloging-in-Publication Data
A catalog record for this book has been requested

ISBN: 978-1-138-72989-6 (hbk)
ISBN: 978-1-315-18959-8 (ebk)

Typeset in Times New Roman
by Apex CoVantage, LLC

Contents

List of illustrations	vii
Acknowledgements	ix

Introduction   1
   Research questions and overview of chapters 1
   Case, methods and ethical considerations 7

1 Slum tourism, subalternity and gentrification   15
   Defining slums and jhuggi jhopris 15
   The agency of the urban Indian governed 18
   The marginalisation of Delhi's jhuggi jhopris 25
   Salaam Baalak Trust's City Walk and the demolition of the Akanksha Colony 30
   Conclusion 35

2 The authentic slum or former street children as prisms of authenticity?   37
   Conceptualising Delhi's informal urbanism as a creative, subaltern space 37
   Discursive and performative approaches to studying tourism 41
   Street life and 'prisms of authenticity' in Pahar Ganj 45
   Conclusion 55

3 Playing with privilege? The ethics of aestheticising the slum   59
   Whiteness and slum tourism 59
   Privilege and playful abjection on the CW 66
   The pedagogical and performative track of the CW 75
   Conclusion 79

4 The affective economy of slum tourism 83
Tourists' responses to the CW 83
Economies of affect and capital in tourism 87
The anxiety of encountering shelter-home children 93
Conclusion 98

5 The post-humanitarian logic of slum tourism 101
Soft- and hardcore poverty porn and ironic humanitarian appeals 101
The anger of encountering shelter-home children 109
The grief and pain of encountering shelter-home children 112
Conclusion 116

6 The emotional labour of CW-guides 121
Collecting data on the CW-guides 121
The shaping of a guide's 'personal story' 123
Excluded stories and ironic performances 129
Subaltern shame and performative therapy 135
Conclusion 139

7 The economy of resocialisation: the slumming researcher? 141
Scripts of involvement and detachment in volunteering 141
My position within SBT 148
The researcher as performer? 155
Conclusion 159

Conclusion and further perspectives 161
The show/shield debate and (im)possible articulations of solidarity 161
Subalternity, meritocracy and hegemony 165

References 169
Appendix 179
Index 181

Illustrations

Figures

0.1	Welcome to SBT – sign at the entrance of the Aasra Shelter Home	7
1.1	Demolition of Akanksha Colony – girl on rubble, captured by Jessie Hodges	16
1.2	Demolition of Akanksha Colony – salvaging water storage utensils, captured by Jessie Hodges	28
1.3	Demolition of Akanksha Colony – salvaging building materials and utensils, captured by Jessie Hodges	29
1.4	Demolition of Akanksha Colony – police and bulldozer, captured by Jessie Hodges	29
1.5	Demolition of Akanksha Colony – salvaging doors, beds and LPG-canisters (gas), captured by Jessie Hodges	30
2.1	Street child collecting trash in Pahar Ganj	45
2.2	The 'God Lane', point (g) of the new CW Route	52
3.1	Pahar Ganj bylane, point (d) of the new CW Route	60
3.2	The trash room near the pottery market on the City Walk	75
4.1	The researcher interacting with SBT-children at Aasra Shelter Home	93
5.1	Interaction between tourists and SBT-children at Aasra Shelter Home	108
5.2	Tourists and SBT-children playing pat-a-cake at Aasra Shelter Home	109
6.1	A CW-guide explains the poster of 'success stories' among former SBT-children at Aasra shelter home at the of the City Walk	123
6.2	A CW-guide expands on the merits of former CW-guides at Aasra shelter home at the of the City Walk	134
7.1	SBT-ambassador posing by a classic car at the Ojas Gallery after SBT's fundraising event	154

Maps

2.1	Map of the old CW-route	179
2.2	Map of the new CW-route	180

Acknowledgements

I would like to thank my mentors, Lars Jensen and Kirsten Holst Petersen, who has spent an inordinate amount of time reading, discussing and advising me. It's been a long road and hopefully it will be longer. Equally warm thanks go out to my other wonderful colleagues at Cultural Encounters, Roskilde University, Denmark. Your work is an inspiration to me.

Everyone at SBT was helpful and welcoming to me and showed a lot of trust in letting me snoop freely around their backyard. Special thanks go out to Danish and Nick, however – you are all over these pages, even if your names are not.

Jan Samuelsen, Rikke Frisk, Cern and Anwesha Chakraborty: your hospitality, friendship and input came at a critical time. Niels and Ida, who are as involved in this project as in my first steps, and Mette and Theo, who keeps me sane. Thanks all.

Introduction

Research questions and overview of chapters

This book studies guided slum tours. Tourists from the global North visit spaces of informal urbanism on the margins of the megapolises of the global South and interact with the subalterns living and working there. In 2015 an estimated million tourists went 'slumming' in this way on the edge of at least 14 cities situated in Latin America, Africa and Asia (Frenzel et al. 2015). There is no reason to think this number has diminished since then, and while it is actually low compared to other forms of tourism to these destinations, such as leisure tourism, it still means that slum tourism quite possibly facilitates the largest amount of organised, face-to-face encounters between subjects from the global North and South. For that reason alone, they are worth studying.

Literature abounds with stories of how representatives from the global North and South physically occupy the same spaces but don't encounter one another. Kiran Desai reminds us in *Inheritance of Loss* (Desai 2006) that representatives of the global South clean, cook, drive and generally maintain the infrastructure of the world's metropolises, and for a meeting to take place between North and South, the dinner guests at a fancy restaurant in Paris, New York or Delhi really only need to seek out the dishwashing station and talk to the underpaid migrants usually working there. However, this does not mean they transgress the stratification separating them in any profound way, and such meetings might be thought of as the accidental by-products of the global division of labour, rather than an attempt to reach across the social divide it creates. Guided slum tours set themselves apart from the preceding in that their explicit theme is the encounter itself and that it physically takes place in spaces that are discursively constructed as the 'home' of the representatives of the global South. Participants of slum tours are promised an insight into the lives of the ethnically and socially defined 'others', which they wouldn't attain if they simply went 'slumming' by themselves. This book examines what this insight might be, whether the promise of an encounter is redeemed and, if so, how.

The book is partly written for an audience of students and scholars interested in new ways of analysing this subgenre of tourism, and the book is therefore divided into chapters that each in their titles signal an intervention in the field of slum tourism studies. Each chapter performs this intervention theoretically, incorporates it into the book's analytical framework and then uses it to analyse different aspects of the cultural encounters taking place within slum tourism. The analytical potential of the interventions is thereby exemplified in each chapter, and readers looking for methodological or theoretical inspiration to approaching slum tourism studies in new ways might thus find the chapter(s) that appeal to them, by reading the overview provided in the following.

The book is also written for scholars and students of other disciplines than tourism studies, such as urban, development, cultural, subaltern, whiteness and postcolonial studies. Each theoretical intervention is performed by reaching out to these neighbouring disciplines in turn so that they might contextualise the analytical results provided by the case of the book, as well as other works of slum tourism research. It is thus the hope that slum tours might become the empirical focus of future studies within these disciplines and thus be analysed as for example a new form of urbanism, as proximate, humanitarian aid performances or as organised encounters between representatives from the global North and South.

The book also works as an ethnography of a group of former street children, who make a living performing 'City Walks' (henceforth CWs) for foreign tourists in the area of Pahar Ganj, Delhi, India. Each year, they show more than 3,000 tourists around on the streets where they used to live before they were contacted by a social worker and came to stay at one of the five shelter homes run by the non-governmental organisation (NGO) Salaam Baalak Trust (henceforth SBT). They run the CW as an education and employment scheme for the guides, who typically stay for three years after they have come of age, after which they ideally move on to other jobs. On the CWs, the tourists are invited to experience the streets where guides used to live and to imagine the lives of the current street children living there now, while they are provided with information about how SBT intervenes in their lives and attempts to alleviate their suffering. At the end of the CW, the tourists are invited to interact with a group of recently 'rescued' street children for 10 to 15 minutes at an SBT shelter home, before the CW-guide tells his or her personal story of how he or she came to live with the NGO. Then the tourists pay, and the CW is over.

Studies of contemporary slum tourism produced within the last 10 years provide ample examples of just how differently the practice is carried out, depending on whether it takes place in Rio de Janeiro (Freire-Medeiros 2009, 2013; Steinbrink 2014) or in South African townships (Rogerson 2004; Koens 2012), among orphans in Cambodia (Reas 2015), recycle workers in Mumbai (Meschkank 2011, 2012; Dyson 2012), or even among voluntourists, who don't think they are slumming at all (Hutnyk 1996; Vodopivec and Jaffe 2011; Crossley 2012b, 2012a). A few books have made a point of comparing the practice in different parts of the world, either by collecting articles (Frenzel et al. 2012; Frenzel and Koens 2014) or in comparative works (Frenzel 2016). This book moves in the opposite direction by focusing on one case but devoting more space to context. The 'field' examined here thereby isn't just the CW as a space of interaction but includes a series of other arenas, where different actors connected to the CW encounter each other. They are guides, tourists, voluntourists, NGO staff, underage orphans and researchers, and each chapter is devoted empirically to such an arena with its own actors and frameworks of interaction.

The book thereby aims to show that the study of slum tourism is connected theoretically to other disciplines because the *practice* of slum tourism is connected to other fields of practice. Urban spaces are created by residents as well as the travellers visiting them, NGOs working in these spaces are sometimes also tour operators, and their visitors might thereby be tourists, volunteers, activists, consumers and donors all at once. Similarly, the poverty performers facilitating these encounters might be guides, subalterns, citizens, labourers and potential receivers of aid. One of the aims of the book is to show how the different groups of actors are strategically framed or frame themselves so that they might achieve certain goals, depending on what they need and want. One such goal might be the visitor's wish to experience certain emotions connected to specific ethical positions vis-à-vis global inequality. Another might be the poverty performer's wish to accumulate different kinds of capital in the co-performed space of affective negotiation constituted by the CW, such as money, knowledge, skills or contacts among foreign volunteers or domestic elites. The fluid identities of these groups and what aspects of those identities they choose to emphasise in certain interactions can only be studied by adopting a theoretically interdisciplinarian framework that devotes ample amount of space to understanding these aspects in their own right, before analysing how they are combined to achieve these goals. That is the reason this book keeps returning to the same case in the different chapters, rather than including some of the many other cases around the world.

Chapter 1 focuses on the question of subaltern agency and the dilemma posed by Partha Chatterjee of how contemporary, urban subalterns might represent themselves if they live in a 'slum' situated on a piece of land occupied paralegally? Should they position themselves as 'citizens' with inalienable rights and thereby risk being displaced as they have no formal claim to the land they inhabit, or should they position themselves as a 'population', within the discourse of governmentality, who has the moral right to the land it occupies, and are thus bound together by a series of subject positions produced in relation to this claim? And if they cannot speak within a discourse intelligible to the elite (or if the elite will not listen simply because it is them speaking) do NGOs facilitating 'slum tours' then have a moral obligation to speak for them? And if so, what choices should they make on their behalf?

Empirically, the chapter provides a view of the recent history of SBT's CW as well as insight into the political climate of the space it exists within. The CW was conducted partially in a small slum called Akanksha, situated within the grounds of New Delhi Railway Station until 2010 when the slum was suddenly demolished by the Delhi Development Authority and the CW was barred from even entering the station. The demolition was a part of a wider move by the civic authorities of Delhi to 'clean up' the city, which posed two mutually exclusive ethical demands of SBT as a socially responsible slum tour operator. On one hand, the need to represent the slums' inhabitants as citizens at par with the inhabitants of the 'official city' was even greater and could be done by inviting foreign tourists to encounter the slum on CWs, where they might also interact with its inhabitants in a respectful manner. On the other hand, this simultaneously exposed these particular slums to the civic authorities of Delhi who were eager to represent the capital as a slum-free space and thus might be even more disposed to demolish slum colonies if they were made into tourist attractions.

Chapter 2 focuses on the idea of the authentic slum and how it might be utilised in tourism performances. It approaches the question theoretically from two angles: one providing examples of how the Indian slum is discursively produced as a desirous, creative, urban space of informalism (Sundaram 2009) and another focusing on how authenticity is continually ascribed, undermined and re-ascribed in representations and performances of tourism destinations (e.g. MacCannell 1976). The chapter then examines how these ideas converge in the discursive production of Pahar Ganj. A touristic borderzone created historically by the influx of travellers, but marked by a version of informal urbanism that caters to certain ideas of transgressive behaviour central to backpackers looking for an anti-tourist experience.

Returning to the case of the book, the chapter analyses how the slum demolition in 2010 forces SBT to change the route of the CW so that it instead leads into the area of Pahar Ganj, which isn't an 'authentic' slum, though some of the guides used to live on its streets before they encountered SBT. The chapter analyses how online material promoting the CW from 2013 focuses on the CW-guides instead of the area, and invite the CW-visitors to replicate their specialised gaze on this otherwise quite ordinary topos. The guides are thus situated as *prisms of authenticity*, who reflect and project an authenticity onto the topos that it doesn't possess in itself. To inhabit this position credibly, they must furthermore position themselves as *liminal subalterns* perpetually on the verge of being resocialised but never quite getting there, as that would undermine their ability to credibly represent the current street children. The chapter ends with an analysis of the performative tools they employ to turn their lack of linguistic and intellectual proficiency into a mark of this authentic liminality, rather than a flaw in their training.

Chapter 3 ties the aesthetics of slumming from the previous chapter (Benjamin and Lacis 1942; Dovey and King 2012) to a genealogy of postcolonial (tourism) studies (Said 1978; Bhabha 1994; Hutnyk 1996), ending in critical whiteness studies and queer phenomenology's focus on the affectively constituted body (Ahmed 2004). It uses this framework to explore the slippage between desire and disgust in slumming and to analyse the temporary, sometimes thrilling, discomfort experienced by slum tourists, as they play with a privilege signalled by the signs on their bodies that stand out in the slum. An approach the chapter dubs *playful abjection*.

The chapter then continues the ethnographic analysis of the CW as a tourist performance in 2013 from the previous chapter, focusing on how the guides' performance of their projected gaze on Pahar Ganj strikes a balance between two 'tracks': a *performative track* that invites the visitors to engage in playful abjection that renegotiates configurations of disgust and desire felt towards street life and a *pedagogical track* that communicates the longitudinal suffering street life exposes the children to over the years. The chapter explores not only how are these tracks are co-performed by CW-guides and -visitors but also how the underlying sensibilities are taught to the guides so that they might facilitate the visitors' experience of them.

Chapter 4 theoretically explores the possibility of an 'affective turn' within tourism studies (Picard and Robinson 2012) and proceeds to conceptualise tourist performances as spaces of negotiation wherein capital and affect circulates in relation to each other in 'affective economies' (Ahmed 2004). In this space, affect is accumulated in the signs circulating among the bodies of the tour's participants, while capital in the shape of money circulates from the tourists to the guides and the organisation they are a part of. Economies of affect and economies of capital are thereby mutually constitutive, as the generated economic capital, in turn, sustains a material framework, which makes the economy of affect possible.

The CW-guides' job is partly defined as creating a space where tourists, who feel de-sensitised by the overwhelming poverty of India, might be temporarily re-sensitised to the plight of street children, an act that is made possible by the donation tourists make to SBT via the tour fee, which works as a tangible proof that they are alleviating the suffering they are presented with. The guides might thereby be seen as facilitators who ensure the circulation of the right kind of affect within the affective economy of the CW, but the chapter proceeds to explore what happens when this goes wrong and anxiety starts circulating between the CW's participants.

Chapter 5 explores theoretically how slum tourism, as a form of pro-poor tourism, is linked to ideas of humanitarianism. In contemporary aid campaigns, donors connect to distant vulnerable or suffering others via aid performances, informed by a new type of 'post-humanitarian' solidarity, where the donors' engagement occupies a playful, ironic space between business and charity (e.g. Chouliaraki 2012). Slum tourists might be said to occupy that same space, the difference being that slum tours are proximate aid performances that bring tourists face-to-face with vulnerable or suffering others. The chapter explores in what way this physical proximity might translate into a metaphorical proximity vis-à-vis the dynamics of compassion? Do the urban subalterns encountered on slum tours remain strangers, or do they come to occupy some other category, and does that change the obligation to help them in certain ways?

The chapter then continues the exploration from the previous chapter regarding the visitors' emotions and asks whether it is possible to transgress the boundaries delimiting the space of comfortable affective negotiation, governed by logics of post-humanitarianism? Is it possible to stray into uncomfortable territory by rejecting the affective dramaturgy of the CW on ethical grounds? How is this ethical discomfort different from the pleasurable discomfort of playful abjection, and why is it important to the guides that the visitors not to feel the former, while the latter feeling is encouraged by them?

Chapter 6 begins by conceptualising CW-guides as emotional labourers, who ensure the circulation of positive affect within the space of the CW. While

Hochschild (1983) warns that alienation might ensue from commodifying workers' emotions and forcibly aligning them with the 'feeling rules' of the workplace, recent research (Johnson 2015) finds that de-commodifying emotional labour has its own pitfalls, as the logic of market exchange provides a possible limit to how emotionally involved workers have to be, precisely because they are defined as being professionally, rather than privately, involved. On the CW, the guides always tell a 'personal story' in the end of how they lost their families and were found by SBT, and comparing the structure of the stories, the chapter concludes they are shaped in relation to a master narrative, whose aim is to support a collective identity among the guides as 'former street children'. This narrative partly situates them as liminal subalterns, perpetually on the verge of complete resocialisation, while simultaneously capable of speaking for current street children, but it also lessens the feeling of stigma associated with illegal, painful or shameful events as they are either omitted or narrated as common to all the guides.

Based on this, the chapter argues that keeping slum tours comfortable by downplaying the suffering of the subalterns they supposedly represent is not only in the interest of the NGOs, which benefit financially, or only the tourists, who get to engage in a playful alleviation of pain. It is also in the interest of the liminal subalterns, who act as representatives of the much larger group of subalterns still living on the streets or in slums, not only because these representatives benefit financially from donations but also because the distance of being professional, emotional labourers helps them overcome the challenge of perpetually having to perform the past suffering associated with being an (almost) resocialised subaltern.

Chapter 7 studies the relation between slum tourism and voluntourism. In studies of the latter (Vodopivec and Jaffe 2011; Crossley 2012b, 2012a), there seems to be a growing apprehension about unequal power relations between actors in the field, which mirrors similar concerns in studies of slum tourism (e.g. Reas 2015). While voluntourism implies a greater commitment to helping than slumming, it is still an open question whether the professionalised, two-hour involvement of paying tourists are more damaging to for example vulnerable children at an orphanage than the two-week or two-month involvement of a gap-year student who inevitably leaves in the end?

The chapter tries to reframe this discussion of exploitation and commodification in voluntourism research to include the types of agency possessed by children involved in these practices. Analytically, it includes a portrayal of SBT as an environment, where SBT-children generally interact not only with a succession of tourists and volunteers but also journalists, potential donors, social workers and researchers like myself. Using an ethnographic account of a fundraising event hosted by SBT as a point of departure, the chapter analyses how SBT-children consciously accumulate social and cultural capital from these encounters in the shape of informal education and useful contacts, but also how the value of these forms of capital are highly contingent on the children continuing to identify with the subject position of 'former street child', because their linguistic, academic and performative abilities count for more if they are seen in relation to their initial disadvantage.

Figure 0.1 Welcome to SBT – sign at the entrance of the Aasra Shelter Home

Issues of exploitation, voyeurism and aestheticism typically studied in slum tourism research thus give way to questions of the long-term effects of for example SBT children spending their entire youth in the company of representatives of the global North, whose social and economic status they can never hope to achieve. The affective economies of the CW are thus theorised as being part of a much larger economy that permeates the relationship between SBT's liminal subalterns and the people who try to help and represent them – including the author of this book.

Case, methods and ethical considerations

The process of selecting a case for this book took place between 2007 and 2011, where I worked as a travel writer for a major Danish publishing house on two guide books on South and North India (Holst and Mukherjee 2009, 2011) and in that capacity travelled India extensively and interacted with its very diverse tourism industry. After having attended the CW in 2009, I decided to conduct a small pilot project in 2011, where I brought a group of postgraduate students from Roskilde University, Denmark to India, made them attend the CW and interviewed them afterwards. Based on this I planned the main part of the fieldwork, which took place during a four-and-a-half-month period between January and May 2013, with a two-week follow-up period in February 2014.

During the initial research trips to India, most of the tourist gazes directed at India that I interacted with invited me to naturalise and accept the juxtaposition of opulence and poverty visible everywhere. A former Maharaja had responded to the forced redistribution of agricultural land in post-independence India by converting his Rajasthani mansion into a luxury hotel, and he spoke at length about the affectionate relationship between the former rulers and their subjects in the nearby village, whose grandchildren still brought him Christmas presents. He discursively produced this quasi-feudal system as an example of an 'indigenous culture' which should be preserved and thereby presented the alignment of me as a tourist with him as a benefactor/landlord as a heritage-preserving act in itself. This invitation to an alignment with a position of 'beneficent ruler' repeated itself in the converted colonial mansions, plantation buildings, hunting lodges and *havelis*,[1] whose owners would encourage tourists to visit the local villages and see the wells being built, the children being taught, the handicapped being cared for, the buildings being restored or the arts and handicraft of the area being reproduced. Ferrying us between hotels and development projects were the drivers in starched shirts who had been trained to call themselves 'travel executives' in order to gain entrance to the lobbies of hotels, which didn't allow 'drivers' inside. By renaming themselves they attempted to trick the implicit class- or cast-based system of 'apartheid' ingrained into the spatial strategies of most tourist places, but they never invited me, or any other tourist I interacted with, to speak up against this system as such.

My initial interest in SBT's CWs was piqued by what I interpreted as a partial departure from this approach. First, the CW-guides represented the suffering and vulnerability they had faced as young rural–urban migrants and/or street children as stemming from relatively complex social problems. Though their political insight and linguistic capabilities were limited because of a lack of education, they didn't try to represent deeply unequal power relations as examples of amicable coexistence between classes, castes and communities, like the deposed Maharaja-cum-hotel-owner described earlier. Nor did they represent SBT as a solely technical or pecuniary fix to social problems, like the well-building projects or the specially erected handicraft-villages, which would allegedly preserve local cultures.[2] As a tourist performance it thus departed from the majority of tourism to the global South that seeks to gloss over inequality rather than to make it an explicit theme of tours.

Second, SBT was successful, both as an NGO and a tour operator. The CW had an ever-growing number of 'visitors',[3] generated an increasing amount of funding for SBT,[4] and it was well represented in reviews in guidebooks and online (TripAdvisor, etc.), as well as in the feedback material filled out by the CW-visitors and collected after each CW by the CW-guides.[5] Seen from the outside, SBT's CW was thereby recognised as a viable example of socially conscious slum tourism, and as I began my fieldwork within the organisation this was largely confirmed by my many interactions with SBT-staff, -children and -guides, and it thereby seemed to live up to international standards of accountability and Indian standards of child care.[6] If there was a case where post-humanitarian slum tourism could generate positive results, then SBT seemed the likeliest example I had come across, and this was important as I wished to analyse the systemic problems connected to this practise, rather than a coincidental instance of mismanagement.

Third, SBT was accessible as a site for fieldwork. During my initial encounter in 2009 it became apparent that SBT was the preferred destination for an ever-changing population of volunteers from the global North, and by inhabiting this position I was not only allowed access to the children, guides and tourists but also entered a relatively well-established position as volunteer within the organisation. Inhabiting this position made my interlocutors respond to me in particular ways and thus influenced the knowledge I was able to produce while there, which is further described in Chapter 7. But even though the tourists on the CW were encouraged to align themselves with the subalterns of India rather than its elite, this alignment was highly temporary and seemed to border on a playful, imagined alignment rather than a politically motivated choice to intervene in the reproduction of inequality that the tourists were so blatantly invited to take part in everywhere else. I therefore set out to analyse how this alignment was discursively produced as a viable option for tourists and with what amount of ease they accepted it.

During the main period of fieldwork in 2013, I spent my days attending a CW in the morning and teaching the guides English and 'presentation skills' in the afternoons, and once a week I taught music to a group of mostly younger children at another SBT-shelter home, in order to get a wider feel for the atmosphere of a part of the organisation that was connected to voluntourism rather than slum tourism. At night I returned to a 'volunteer flat' tucked away in a small alley in Pahar Ganj provided by SBT and run by former CW-guides, whom I hung out with, along with the changing population of volunteers housed in the flat.

My approach to data collection was multifaceted and evolved gradually. I built relations of trust within the organisation, gathered material and developed increasingly precise and context-specific research questions. By the end of the fieldwork my data consisted of written material, observation notes, interviews and recordings of CWs, training sessions and other events. As I began to analyse the data I had collected, some of it came to function directly as empirical material to be analysed in these pages, while other parts were used purely as background material.

The written material that I produced and collected during my fieldwork consists of a CW-script, a map of the CW's different routes coupled with extensive observation notes of what took place on the CW and a year's worth of feedback forms about the CW produced by SBT for internal use. It turned out that the former administrators of the CW had also produced texts about the CW, and I collected to two internal reports about the progress of the CW from previous years and the 111-page *City Walk: Guide Training Manual* (Hodges 2010) compiled by a former CW-coordinator. Observation notes were collected in a diary I kept throughout my stay, and as I began writing the articles the diary came to act as a repository of context that I would consult whenever the documents I had gathered or the interviews I had conducted were inconclusive in themselves.

The interviews were conducted in circumstances that differed significantly from each other. I had chosen to remain with SBT during my fieldwork, rather than follow a group of tourists around, and I therefore had limited access to the tourists, who were usually on their way to another attraction right after the CW. I would find my interlocutors among the tourists by observing a CW and then raise my request for an interview right after they were finished, either at a nearby café or whenever they had time before they left Delhi. The underlying inquiry that I pursued throughout these interviews was a delineation of the boundaries surrounding what the tourists perceived as a space of comfort on the CW, but as I had no privileged access to those emotions, I had to contend with the fact that the reactions I sensed in them might be wildly inaccurate. I therefore resolved to let my questions be directed by the behaviour I thought I had observed in them during the CW and so simply ask whether what I had sensed in them was correct or not. Rather than designing a specific interview guide that I submitted all my interlocutors to (see e.g. Kvale 1996), I therefore shaped my questions in a dialectic relationship between their reactions, as I had perceived them on the CW, and my theoretical framework, which evolved gradually the longer I stayed with SBT. The results of this are presented in Chapters 4 and 5.

The data collected from my interactions with SBT-children, -adolescents and -staff were much more varied, as my relationships with them had more time to evolve. While I collected the written version of the CW-script and mapped the CW's route, I also initially recorded a number of CWs both by audio and by video, and while this was encouraged by the guides, who wanted to improve their English in subsequent lessons based on this material, it did prevent me from focusing fully on the tourists' reactions, and so I had to choose whether I wanted to focus on the guides or the tourists when going on a CW. I quickly found out that each guide had a set format of how to conduct the CW, which neither differed significantly between performances nor even between the guides, as each guide used the same script. What stood out, however, were their performances of a 'personal story' at the end of each CW, and a series of recordings of these came to act as empirical material of Chapter 6.

The analysis of the personal stories is contextualised by a series of interviews I conducted with the guides, the current and former CW-coordinator and a British volunteer, who had been involved with the CW longer than anyone else working with it. The interviews were conducted towards the end of the main period of fieldwork in 2013, because I needed time for my interlocutors to open up to me and for me to know what questions to ask, and in this case I was much closer to employing an interview technique akin to that of Kvale (1996), where the interview is preceded by so much research that the questions are certain to be relevant to the topic the interviewer wants to explore. I developed an interview guide to be used for the interviews with the CW-guides in collaboration with the current SBT-coordinator and the aforementioned British volunteer, and shaped it as an evaluation of the CW seen from the perspective of the CW-guides. In the formulation of the questions I made sure that their answers could be used both to contextualise my inquiries about what it meant to be a CW-guide and the SBT-staff's inquiries about how the guides thought the CW might be improved seen from their perspective. I submitted my findings to the SBT-coordinator in the shape of an evaluation of the CW as a workplace, where I made sure that critical views of the management of the CW were anonymised to the extent that I could. This evaluation then came to act as the interview guide for my interview with the SBT-coordinator where she was given a chance to reflect further on the guides' answers and contextualise them.

Conducting my fieldwork within SBT raised a series of ethical questions of how consent might be sought from the subjects I wanted to participate in my study. The staff stipulated that I could not interview anyone younger than the age of 18, and as all guides were of age that was not a problem, though special permission had to be made for the 17-year-old trainee guides, who gave their explicit consent to participate. The paradoxical context of this stipulation was that SBT-children as young as 6 years old were interacting with more or less randomly chosen outsiders on a daily basis and were for example put in the immediate care of 21-year-old volunteers, who would usually leave after a month to be replaced by other volunteers. Whether this had detrimental effects is hard to say without a thorough psychiatric evaluation of the children in question, but it could in any case not be viewed separately from the context SBT operated within, with under-staffed shelter homes where they needed all the help they could get.

Granting me access to underage SBT-children while banning me from representing this interaction in anonymised studies could be interpreted as though SBT was less concerned about how they cared for their charges than about how this was represented to the media, but this was only true from a certain perspective. To the staff of SBT, the potentially harmful effects of being exposed to volunteers, tourists, donors, journalists or even researchers like me had to be weighed against what might happen if the children were left to themselves. Had SBT received more funding, it would not have had to use gap-year travellers to care for vulnerable children or expose them to tourists or donors, but as it received next to nothing from the Indian state, SBT had to rely financially on these activities. It was aware, however, that this might easily be portrayed in the media as a way of prostituting the misery of their charges, because an article in the British newspaper *The Guardian* had already implied as much (Gentleman 2006). Aware of how important it was to frame the encounter between SBT-children and outsiders as respectful, the CW-coordinator even discouraged me from employing labels such as 'slum tourism' or 'poverty tourism' as she thought they gave the wrong connotations. As the activity they facilitated on the CW had all the significant characteristics of these types of tourism, I couldn't grant them this wish but resolved instead to explicate how SBT facilitated its performance in a thoroughly responsible manner within its context, and hoped this would save them from the unproductive, viral outrage they so wanted to avoid.

Furthermore, the names of all SBT-children and -adolescents have been changed. Even though the personal stories of the guides are made public every time it is told to a group of CW-visitors, the CW-guides have still been robbed somewhat of the power to represent themselves in this dissertation as I reserve the right to frame their stories theoretically in ways they might not agree with. In the analyses, I raise the question of whether an almost daily articulation of a potentially traumatic past is necessarily healthy, and it therefore seemed prudent not to let this book be a permanent record that their stories were, in fact, theirs, though I am sure they wouldn't have minded if I had asked them at the time of telling to appear in these pages with their names. I also anonymised the tourists I interviewed, partly because it gave me the same freedom to somewhat reinterpret their reactions within a theoretical framework they were not aware that I would use during the interview but mostly because I was looking for emotional reactions among them and thus represent them at their most vulnerable. For the same reason, I was hesitant to include pictures of the CW-visitors, CW-guides and the SBT-children in this book. Chapter 3 provides an account among the SBT-staff and volunteers about whether CW-visitors should be allowed to take pictures of and with the underage SBT-children on the CW, where parts of the SBT-staff thought it demeaning and unsafe that these pictures entered social media accounts of visitors, without the children having any say in how they were represented. On the other hand, the SBT website showed lots of pictures with happy SBT-children playing, and by the end of my fieldwork, where my role had gone from being a tourist to being halfway part of the SBT-staff, they even asked me for pictures of the CW to put on it. I struck a balance by asking permission from everyone I took pictures of and by reminding them at intervals that I was writing what might become a book one day and that they might go into that.

Another ethical consideration was whether to represent the way both guides and CW-visitors spoke English with all the errors of grammar, syntax and pronunciation of the non-native speaker? Of the fourteen interviews I conducted with tourists during my fieldwork I quote three, but two of these three are conducted in Danish and translated by me. This leaves one non-native English-speaking tourist, and her creative use of a word ended up providing an important analytical point. A more serious consideration was the CW-guides, who could easily come across as less intelligent than they, in fact, were if only quoted when speaking English, which wasn't their mother tongue. Transcribing their flawed English, I was reminded of Edward Said's (Said 1993, 19–31) problematisation of Conrad's (1902) *Heart of Darkness*, where Conrad let Africa be the backdrop to the white protagonist's journey into the insanity of colonialism with only sporadic lines of incorrect English being articulated by the 'natives', like *'Kurtz. He dead'*. In a contemporary setting, Khair (2000) analyses the proliferation of 'Rushdie-English'[7] in Indian writing, where English is peppered with expressions in Indian languages (mostly Hindi) to give readers with no proficiency in these languages a vaguely 'Indian feel'. Khair points out that this overshadows the fact that India is overwhelmingly bi- or trilingual, and that most conversations within the social context that Rushdie portrays would be carried out in two or more correctly spoken languages. In both cases the (post)colonial other is caught at a disadvantage, and true difference is erased, in favour of an intelligible fantasy of this other.[8]

When my fieldwork ended, my recorded material featuring the CW-guides consisted almost exclusively of them speaking English, either to tourists or to me during interactions where they wanted to practise their English. In this book, this overshadows the fact that many of them were quite eloquent in Hindi and that their conversations outside SBT was overwhelmingly bilingual, or even trilingual if they had come from areas where Hindi was not a typical first language, like Assam or enclaves in Bihar or Uttar Pradesh, where local languages or dialects were spoken that differed significantly from Hindi. Rather than correcting what they said to represent their true linguistic abilities in these pages, however, I chose to represent the interaction on the CW as it took place linguistically, along with all the possibilities of the participants misunderstanding each other because of a lack of linguistic abilities, as the insecurity this caused affected how emotions circulated and were amplified among the participants of the CW.

Finally, there are the questions of what relationships I as a researcher entered into with my interlocutors and how I, along with everyone else within the field, was guilty of reproducing the very structures of inequality I was critiquing in the first place. Rather than apologising for this fact, I made it part of the topic of Chapter 7, where I analyse how the affective economy of the CW extends into other relationships than the one between guides and tourists.

Notes

1 Rajasthani mansions.
2 I am here thinking of for example Raghurajpur Artist Village in Orissa or Choki Dhani in Rajasthan, or perhaps even Dilli Haat in Delhi.
3 This was the word used within SBT to signify the tourists going on the CW.
4 This is substantiated by internal reports I gained access to during my fieldwork, as well as the balance sheets published every year on SBT's website. The relative modest amount of donations from 2013 (CW-visitors times the mandatory CW-donation, 3,000 × ₹200 = ₹600,000 [$9,000]) is increased by the fact that many visitors donate more than the mandatory amount, combined with the estimate that the CW, as the international face of SBT, generates donations that cannot be linked directly to it.
5 See Chapter 4.
6 According to the National Commission for Protection of Child Rights, see ncpcr.gov.in.
7 A reference to the popular and controversial writer Salman Rushdie, who for a time seemed to dominate the field of literary representations of India in the global, anglophone North.
8 A fuller account of this discussion pertaining specifically to Indian writing in English can be found in for example Mukherjee (2000).

1 Slum tourism, subalternity and gentrification

Defining *slums* and *jhuggi jhopris*

What is a slum? This apparently simple question is linked to a more intricate one. What are the limits of the knowledge it is practically and ethically viable to articulate about 'slums' and the urban subalterns inhabiting them? The pronounced goal of most slum tourism is a respectful encounter between the global North and South in spaces that are discursively constructed as the 'home' of the representatives of the global South. But what are the implications of representing slums? What happens for instance if their added visibility increases the chances of them being cleared by bulldozers? Does this threat influence how we might produce knowledge about slums? Or whether we even use the word?

Mike Davis's book *Planet of Slums* cites the first 'scientific' survey of slums in America *The Slums of Baltimore, Chicago, New York and Philadelphia* (1894) to '*define an area of slum as "an area of dirty, black streets, especially when inhabited by a squalid and criminal* population"' (Wright 1864, 11–15 in Davis 2006, 22, my italics), thus indicating that the squalor of slum areas is reflected in the 'squalor' and 'criminality' of the people living there, as well as notions of 'race'. Roughly 20 years before however, in 1872, Friedrich Engels writing on *The Housing Question* had quite another assessment of why areas of squalid working-class housing could not be abolished:

> The breeding places of disease, the infamous holes and cellars in which the capitalist mode of production confines our workers night after night, are not abolished; they are merely shifted elsewhere! The same economic necessity that produced them in the first place, produces them in the next place.
>
> (Engels 1872, 77)

To Engels, the 'infamous holes and cellars' are indeed a mark of shame to the metropolis, but not one that should be borne by the people inhabiting them but rather by the bourgeoisie, whose capitalist mode of production apparently necessitates the existence of these areas.

Seeking what one might call a de-politicised midway between the two views, the *Oxford English Dictionary* (*OED*) roughly 100 years later defines a 'slum' as 'a squalid and overcrowded urban area inhabited by very poor people. A house in

16 *Slum tourism, subalternity & genrification*

Figure 1.1 Demolition of Akanksha Colony – girl on rubble, captured by Jessie Hodges

such a place'. Wright's and the *OED*'s definition both use the word *squalid*, but whereas Wright uses it to describe the *population* of a slum, the *OED* uses it to describe the *area* where they live, which is a marked departure from his approach. On the other hand, the *OED* avoids commenting on how the 'poor people' that inhabit slums become and stay poor, thus showing no partiality towards Engels's view either. Similarly, Davis notes that other definitions of the word *slum*, like the one adopted by the United Nations (UN) at a conference in Nairobi in 2002, 'discard [Wright's] Victorian calumnies [but] otherwise preserves the classical definition of a slum, characterized by overcrowding, poor or informal housing, inadequate access to safe water and sanitation, and insecurity of tenure' (Davis 2006, 22–3). Still, it is not the case that these dispassionate definitions have completely supplanted the politically charged connotations of the earlier age and when the question of, say, the clearing, remodelling or developing of slums arises in public discussions, one might easily detect a semantic slippage between the word connoting something *depraved* (Wright), something merely *poor* and *underdeveloped* (UN and the *OED*) or something *deprived* (Engels). As we shall see, this is also very much the case in the Indian metropolis.

When the noun *slum* is turned into a verb in the expressions 'to slum' or 'to go slumming', both have a pejorative ring to it. The *OED* defines it as 'voluntarily spend time in uncomfortable conditions or at a lower social level than one's own', so 'to slum', according to the *OED*, does not necessarily mean to visit an actual slum but rather to act like an inverted snob, by seeking lower standards of living than is strictly necessary. However, visiting actual slums for recreational purposes is by no means a new practice. Seth Koven's *Slumming: Sexual and Social Politics in Victorian London* shows for instance that the then slum districts of Whitechapel and Shoreditch were mentioned in the 1887 edition of Baekeder's guidebook of London as places one could visit (Koven 2004, 1, 244). But even while visiting the slums, the Victorian upper and middle classes' conception of it was ambivalent – a messy 'mingling of good intentions and blinkered prejudices that informed their vision of the poor and themselves' (2004, 3), and the 'slum' was thereby simultaneously an exoticised place of abandon, where the presence of for example 'opium-dens' and 'whorehouses' signalled a loosening of the moral strictures imposed by Victorian public life but, therefore, also a place the 'comfortable classes' felt compelled to reform. The 'slum' was therefore not only simultaneously depraved and deprived; it was also strangely *desirous* – partly because it was *dangerous* and vice versa.

As Dovey and King remind us in a contemporary setting (2012) the desire towards the slum might take the form of a fascination not only with filth, squalor and projected moral licentiousness, but also with the 'sublime' juxtaposition of opposites, that is the act of passing from the official city into its seedy underbelly, or perhaps even being able to hold both cityscapes in a single gaze simultaneously. The perceived danger that is linked to this desire therefore lies not only in what might befall the unwary tourist visiting such a seemingly lawless place but also in the moral implications of projecting these desires onto the perceived misery of others. See also Chapter 3.

Academically and ethically it is problematic to define what a 'slum' is in common parlance, without taking into account this very long history of sociopolitical struggle inherent in the sign, which (as we learn from Koven) is also inscribed into fantasies of both 'helping those in need' and 'reaching across the divide' into something unknown. As we move into modern-day slum tourism, Steinbrink (2012) argues that we have gone from 'moral slumming', referencing Koven, to 'ethnic slumming', exemplified by historic accounts of trips to ethnic settlements in New York, to 'global slumming', where tourists from the global North travels to sites of poverty in the global South, where they by virtue of their mobility construct themselves as 'global travellers' as opposed to the 'locals' they are there to meet. Commenting on this, Frenzel (2012, 60–1) suggests that the perceived justification for this global slumming among politically conscious visitors from the global North can imply an 'Othering' of the slums' inhabitants as well as a 'same-ing'. 'Othering' would imply that the poor, southern 'locals' are valued others, whose experience might teach the 'globalised' visitors important lessons, whereas 'same-ing' attempts to frame them as participants in the same global, economic system of exploitation that the visitors are also trying to combat, thus making them 'comrades in arms' of a sort. As examples of this practice in an Indian context we can turn to the studies of Reality Tours' work in Dharavi, Mumbai (Meschkank 2012; Dyson 2012; Ma 2010). In their different ways all these studies support Frenzel's impression that most tourists generally seek to reach across to a valued other or same in the slum and that this wish probably extends beyond politically conscious tourists to include most visitors. The question is, To what extent it is possible for them to do so?

In studies of how the Indian elite relates to the slum and its inhabitants in public discourse a quite different picture emerges. There are plenty of examples of 'slums', or *jhuggi jhopris* in Hindi, being defined negatively as that which is *not* affluent, *not* ordered, *not* modern and, therefore, should *not* possess a legitimate legal status. Neither Wright's nor Engel's depart significantly from this reasoning of othering as Wright discursively produces the slum and its inhabitants as the other of the modern nation and its metropolises, whereas Engels claims that they are other*ed*, that is to say forced to live as anomalies within the nation/metropolis. Accounts of historical slumming from Koven and Steinbrink likewise imply the same othering, and it, therefore, does not seem tenable to use these as the theoretical basis of a conceptual framework that seeks to escape the image of the slum as a negatively defined category. One might think that it would be possible to use the lessons learned from globalised slum tours to do so, but as we move to Frenzel's analysis of the tourists' attempt to embrace the slum as a valued other or same, both justifications of slum tours seem to gloss over the complicity of the slumming visitors. The othering of them tries to confer value onto a perceived difference inscribed by poverty, but one might argue that this really only changes the stereotype of the slum and its inhabitants to a positive one rather than a negative one. Similarly, the same-ing of the slum simply erases the difference inscribed by poverty in the name of solidarity, as well as the fact that the tourists in a number of ways help to perpetuate the economic system that reproduces the inequality they purport to fight.

The agency of the urban Indian governed

But what about the knowledge about slums produced by inhabitants of slums? Here, we run into the conceptual problems articulated by Spivak (1988) about the possibility of the subaltern to speak, be heard and be understood within public discourse. This is not just a question of whether they are given time in the media to represent themselves or of whether the representatives chosen within the media are truly representing all of them or of whether they speak the language of money and power (English or perhaps Hindi) and with what proficiency? It is also a question of what knowledge they are able to produce about themselves and the places they live, within the discourses made available to them by the elite.

The genealogy of the contemporary, urban subaltern arguably starts with Antonio Gramsci's critique of Marxian materialism in his analysis of *The Southern Question* (2005 [1926]), where he explores the possibility of the peasantry of the south of Italy uniting with proletarian workers from the north against the exploitative landowners and owners of the means of production in Italy at large. He writes in opposition to an understanding of the peasants as a class of people who have not yet reached political maturity, and who needs the insight of the industrial workers in the North to be able to organise themselves in the struggle against exploitation and that they are necessarily natural allies in this respect. Gramsci sets out to analyse the cultural and political setting of the south in order to frame the choices of the peasants as political choices, rather than actions brought about by false consciousness or political immaturity.

These thoughts are developed theoretically in his *Prison Notebooks* written 1926–37 (Gramsci 1971 or Gramsci 2011). Here, he argues that hegemony is created by a type of coercion that works not just through blunt force but also via consent, which is ingrained into the subaltern subject by public institutions such as churches and schools that among other things create structures of incentive that encourages this subject to advance within the social systems created by these institutions. This double-edged form of coercion is both exerted and resisted, and hegemony can thereby be defined not as an absolute form of control but rather as an unstable equilibrium of compromise (Gramsci 2011, 508–9).

A generation of Indian intellectuals has elaborated on this framework from the Kolkata-based Subaltern Studies Group to later incarnations in Delhi at Jawaharlal Nehru University or the Centre for the Study of Developing Societies. This book identifies Partha Chatterjee (1986, 2004, 2013a, 2013b) as an overarching figure that spans this history of theorisation. In his writings, he connects the historical project of the 1980s of tracing the absence of the subaltern subject in colonial and postcolonial archives, to the ongoing project of exploring the discursive space of agency granted the contemporary 'governed', who might be positioned as subalterns within particular social settings, and his thoughts are thus pivotal to the analysis of the space of agency allowed the contemporary, Indian, urban subaltern.

Subaltern studies in India began as a way of writing Indian history from a perspective that departed both from colonialist elitism and an Indian bourgeoisie-nationalist elitism that, according to Partha Chatterjee, permeated the environment of South Asian history in the 1970s. In his essay 'Brief History of *Subaltern Studies*' he describes two competing approaches, where one based in Cambridge

> argued that Indian nationalism was a bid for power by a handful of Indian elites who used the traditional bonds of caste and communal ties to mobilize the masses against British rule.
>
> (Chatterjee 2010a, 291)

Thus the university-educated leaders of the Congress Party and the Muslim League are cast in the role of seducers of the illiterate masses. The other approach among Delhi-based historians

> spoke of how the material conditions of colonial exploitation created the ground for an alliance of the different classes in Indian society and how a nationalist leadership inspired and organised the masses to join the struggle for national freedom.
>
> (Ibid.)

That is to say that an oppressive colonial rule paved the way for nationalist leaders. Neither of these two historiographical traditions based on the logics of imperialism and nationalism respectively 'had any place for the independent political actions of the subaltern classes' (Ibid.). Drawing on Gramsci's work on how the peasants of southern Italy might be seen as a politically conscious class, whose choices should be read in particular cultural settings, the Subaltern Studies Group set out to write history that recognised the Indian subaltern as a political force – not least in the struggle for independence.

Gramsci's approach provides a conceptual way into exploring subaltern resistance in colonial India, but as subalterns are defined as a class of people situated outside – though dominated by – the hegemonic power structure, trying to understand them on their own terms is conceptually difficult, and this is illustrated by Chatterjee's own study of 'The Colonial State and Peasant Resistance in Bengal, 1920–1947' (Chatterjee 2010a, 302–40). Chatterjee is, in a sense, forced to represent subalterns through the lens of resistance because they only appear in historical records kept in colonial India when they rebel against the colonial state, and this also means that they are always mediated to the researcher through the eyes of the colonial administration's chroniclers. Methodologically, Chatterjee's study therefore consists of taking historical documents, which were most often 'prepared by official functionaries' and then reading them 'from the opposite standpoint' (Chatterjee 2010a, 292), and the voice of the historical subaltern in this study is thus mostly reconstructed from sources that are hostile to them and only mentions them when they 'make trouble'. Chatterjee is thereby left with empirical material stating how subaltern rebellions have been refracted by its chroniclers, and the nearest he can come to emulating Gramsci's project in *The Southern Question* is to guess what gaps, silences and inconsistencies in the archives mean.

This methodological conundrum is further highlighted in the works of another member of the Subaltern Studies Group, Gyanendra Pandey, whose studies of communalism in colonial India (1989, 2006) use historical accounts of communal riots as empirical material. From this, it cannot be conclusively determined what some of the riots were actually about, because accounts of the chains of events leading up to the riots often contradict one another, but what can be determined is that certain early accounts of riots came to act as Master Narrative for later accounts. In this Master Narrative so-called sectarian 'natives' ('Hindoos', Muslims, Sikhs) apparently clash with other 'natives' after which the colonial power intervenes and peace is restored – often by quite violent means, though this is toned down in the reports using this narrative model. And though this master narrative is challenged by other sources claiming that at least some of the subsequent riots were directed at the colonial administration itself, rather than at other 'natives', the master narrative still act as a matrix for the writing of subsequent histories (2006, 16). The study provides an example of how subaltern resistance is written by colonial historians as featuring 'natives', who are represented as fighting amongst themselves, when evidence would suggest that they were *also* fighting against the colonial state and not always within the framework of sectarian communalism, which was seen by the colonial state as an example of pre-modern allegiances to religious faiths rather than to the state. In agreement with Chatterjee's assessment of Indian bourgeois-nationalist historians in the 1970s, Pandey concludes that the Master Narrative instated by the colonial administration was surprisingly readily reproduced by Indian historians when chronicling riots in independent India (2006, 20–2), and as we move on to Chatterjee's analysis of the use of identity politics in 'political society' today, we see similar dynamics at play between religious minorities and the nation, where competing cultural identities, based on religious community, caste or class are discursively produced as creating separatist sentiments among populations that are perceived as subverting the nation's unity.

In response to the Subaltern Studies Group's search for a subaltern subject, which is only traceable in the colonial records via its apparent absence, Spivak pinpoints a series of contradictions in the Subaltern Studies project in an essay written for the group titled '*Can the Subaltern Speak?*' (1988). She is doubtful that this is the case, and as an illustration she describes how widows in 19th-century India were caught between an interpretation of Hindu scriptures that apparently demand their immolation on their husband's funeral pyre and laws imposed by the British colonial force banning this practice in 1829, thus positioning the widows between two 'dialectically interlocking sentences that are constructible as "White men are saving brown women from brown men" and "the women wanted to die"' (1988, 93). She argues that though the widow can thereby be traced in the pages of history, neither of these two dominant discourses leave space for her to actually 'speak' to us from them. The productive outcome of Spivak's intervention is an increased attention paid to the conditions of representation that her other writings develop further (Spivak 1987, 1993, 2003) and which come to influence Chatterjee's writings in 2004 on whether the contemporary, urban subalterns living within the informal urbanism of India's metropolises can speak and, if so, within which discourses.

To situate this within a contemporary frame, a short genealogy of postcolonial theory and its influence on Subaltern Studies must also be traced. Around the time when the Subaltern Studies Group starts publishing from Calcutta, Edward Said publishes *Orientalism* (1978), where he methodologically combines Foucault's (1972) use of 'discourse' as a way of tracing genealogies, and thereby identify recurring ideas through time, with Gramsci's focus on the subject's space of agency to act within the structures created by these discourses and thereby to gradually change them.[1] Homi Bhabha's (1994, 72–3) critique of Said's lack of focus on the colonised subject's inherent ambivalence in its articulation of identity was addressed in the afterword to the later edition *Orientalism* (1994), as well as in his focus on 'counterpoint' as a strategy of reading texts (Said 1993), which later became a strategy for understanding the world at large, as well as its subjects[2] (Barenboim and Said 2002; Said 2006; Holst 2012). However, Bhabha's focus on the instability of the subject and the nation it imagines itself to inhabit and thereby creates informs Chatterjee as he attempts to move on from the theoretical impasse articulated by the early Spivak.

As Chatterjee makes the transition from writing about rural, historical subalterns to writing about *The Politics of the Governed* (2004) in a largely urban, contemporary setting, he not only dedicates the book to Said but in a sense returns to Gramsci's project in *The Southern Question* of analysing the boundaries of the space of agency allowed a subaltern subject, who is *not* displaced in time and thus only analysable through the absences, silences and distortions of colonial archives. So whereas The Subaltern Studies Group based its work on a rejection of the notion 'that the peasantry lived in some pre-political state of collective action' (Chatterjee 2004, 39), his later project is to understand the relatively recent 'entanglement of elite and subaltern politics' (Ibid.). His conclusion is that even though the lives of the subaltern classes have been brought under the influence of the democratic process in India since independence, this does not mean that they have been granted citizenship in any real sense but largely remain what he terms a 'population'.

Chatterjee arrives at this conclusion by critiquing Benedict Anderson's understanding of 'bound' or 'unbound' *serialities* (Anderson 1998) as different ways the subject might articulate its sense of belonging to imagined communities:

> One is the unbound seriality of the everyday universals of modern social thought: nations, citizens, revolutionaries, bureaucrats, workers, intellectuals, and so on. The other is the bound seriality of governmentality: the finite totals of enumerable classes of population produced by the modern census and the modern electoral systems.
>
> (Chatterjee 2013a, 6)

Whereas unbound serialities are not usually mutually exclusive, as the subject can well be a 'worker' as well as a 'citizen', the components of bound serialities are apparently constructed as being mutually exclusive so that it is seen as impossible to belong to several religions, ethnicities, genders and so on. If the subject chooses to define itself in terms of unbound serialities, and thereby refrain from aligning itself with exclusionary categories, it might thus presumably keep itself in a state of fluid, constant becoming, which is a state less prone to inhabiting antagonistic, sectarian and/or 'communal' identity formations.

Chatterjee's response to this idea partly consists of a historical critique of the idea of universal civil rights as the protector of these unbound subjects, and he especially examines the origins of these rights as instruments of domination towards categories of colonial subjects, who were not included in them and thereby could not be protected by them.[3] The second part of his critique focuses on present-day groups like squatters, who cannot sustain their livelihoods by referring to civil rights. They might therefore be compelled to represent themselves as 'population groups' defined by 'bound serialities' in order to get access to special quotas or favours from representatives of political parties, instead of the unbound serialities that grant them none of that.

While adopting Anderson's split between citizens and population groups, Chatterjee also critiques it by reconceptualising the time/space of the nation. The image of the nation as a coherent entity moving through calendrical time as a train on a track has been described by Anderson – with the help of Walter Benjamin – as existing in 'empty homogeneous time' (Anderson 1991, 25), a temporality which, as opposed to messianic time, has no apocalypse at the end of it and is therefore 'empty', but also a temporality which all the members of the nation are imagined to inhabit simultaneously. It is an image of the nation as an interconnected web of relations that no individual can encounter in its totality as anything but an *imagined community*, which is 'homogeneous' because it encompasses all the members of the nation who share this national identity. Chatterjee renames it the 'time of capital' and claims that it is deployed by the elite as the *only* space/time of the nation, and consequently 'when it encounters an impediment, it thinks it has encountered another time – something out of pre-capital, something that belongs to the pre-modern' (Chatterjee 2004, 5). Certain technologies, modes of production, ideas, practices and ultimately subjects – especially those who position themselves in relation to what Anderson calls 'bound serialities' – are thereby discursively produced as belonging to a negatively designated historical past, which should have been left behind but somehow hasn't.

Chatterjee's states that it is the very category of the empty, homogeneous space/time of the nation that excludes them, by discursively producing this image of the nation (that they cannot hope to inhabit) as the *only* one.[4] As an alternative, Chatterjee suggests that the nation must be seen to exist in 'heterogeneous time' (Chatterjee 2004, 7), where different perceptions of the 'now' that the nation inhabits exists side by side but where certain dominant conceptions of 'the modern' are deployed for political reasons. To conceptualise this deployment, he draws on Homi Bhabha's idea of the 'double-time' of the nation, where 'the people' is simultaneously an object of pedagogy, understood as a collection of subjects who are never fully developed as citizens, and an eternal category, whose allegiance to the nation must be performed continuously in order to perform the nation itself (Bhabha 1990). Based on Anderson's distinction between subject positions that might be bound in serialities or not, Chatterjee invokes a similar distinction between two conceptions of 'the people', which are (1) a 'population' who needs to be monitored and governed for their own good but also (2) 'citizens' who *are* the nation and decides through representational democracy how the nation should be run (Chatterjee 2004, 34).

Within this framework, the subaltern is discursively produced as belonging to a negatively designated historical past that is emphatically *not* modern, and it places them within the category of 'population', not 'citizen', thereby essentially turning them into a problem to be solved by the nation, rather than what *constitutes* the nation and its metropolises. There might, however, also be incentives for subjects positioned by the elite as pre-modern subalterns to internalise this discourse and thereby articulate their subject positions as aligned (or bound in serialities) under the banners of caste, class, community, gender and sexual orientation, among others, in order to gain influence within political systems.

One of the most well-known examples of this in an Indian context is how caste identities have become entangled with politics. Chatterjee reminds us of Ambedkar's[5] assertion in the 1930s that Dalits ('untouchables') should have special access to education and government jobs, as he believed that representative democracy in independent India would not provide equality between the castes. Whereas Gandhi insisted that Hindus were an ultimately homogeneous group consisting of upper and lower castes who should be treated equally, Ambedkar did not believe that the incorporation of Dalits into this homogeneity could occur if the elite did not also see themselves as included in the pedagogical project of creating a nation of equals, and thus, he insisted that Hindus were a heterogeneous group who should have access to the state via separate channels (Chatterjee 2004, 13–17).

Consequently, 'Scheduled Castes', 'Scheduled Tribes' and 'Other Backward Classes' were given exclusive access to a certain portion of government jobs and slots in public universities and in 1980 the Mandal Commission raised the percentage of the population included in these categories from 27% of the population to roughly 50%. Since then the question of whether these should be preserved, and if so why, has sparked not only debate but riots as well. Arguments *for* their preservation utilise discourses of separateness from mainstream society, partly due to shared cultural identities within the groups, partly due to outside pressure in the shape of upper caste oppression, whereas arguments against utilise discourses of meritocracy, reminiscent of those employed in arguments against affirmative action in the global North, but also discourses that construct ethnic, communal or caste-based identity formations as pre-modern, especially when they are seen to provide 'vote-banks' for politicians (see Dirks 2001, 265–96; Chatterjee 2010b; or for a popularised version Gupta 2000, 117–45).

An illustration that these types of identity formations are sometimes embraced out of necessity is provided by previous studies of squatter's associations in Kolkata, which Chatterjee analyses within his theoretical framework, and he concludes that they claim a common identity,[6] move from a discourse of civic rights to a 'moral rhetoric' (2004, 60) of how they as a population group should be given exceptional entitlements to tenure on the land they occupy in a state of paralegality and then make alliances with political spokespersons, who might benefit from granting them this as an exception.

The agency possessed by contemporary, urban subalterns is thus to act within political society as part of a group, bound together by a series of subject positions framed within the discourse of governmentality, which they inhabit not because they necessarily see these as more central to their identity formation than those formulated within civil society, but rather because they gain larger political influence by doing so, even if they thereby leave themselves open to allegations of engaging in 'pre-modern' identity politics.

As we shall see, this also has a profound effect on how urban subalterns represent themselves in the encounter with non-governmental organisations (NGOs), foreign tourists and a donating elite and thus also how it is possible for the researcher to examine this exchange, as it is situated within discourses some of which might not even be visible to the subjects interacting.

The marginalisation of Delhi's jhuggi jhopris

The preceding account provides the reader with some historical context to the sociopolitical struggle inherent in the sign 'slum' and how naming an area a 'slum' can have profound implications for the fate of the area and its inhabitants. This is then linked to the space of agency ascribed conceptually to the contemporary, urban, Indian subaltern inhabiting these spaces of informal urbanism that might be named 'slums'.

The next section moves to the space-time of Delhi in the 2000s, which is the empirical and analytical focus of this book, where there are many examples of how discursive battles over the conceptualisation of slums and their inhabitants are linked to a widespread practice of slum-clearances. On one hand, this puts the knowledge production of this book into perspective and illustrates a basic condition it has had to take into account. The section, however, also tries to chronicle how this culture of slum clearances resulted in an intellectual counter-culture celebrating the creative potential of informal urbanism (slum), a potential that might (perhaps inadvertently) give us an initial sign of what alluring sights the Delhi slum might hold if visited by foreign tourists.

One of Chatterjee's points noted in the preceding section is that when urban subalterns strategically align themselves with certain population groups that need to be governed for their own good, it is less proof of their inherent pre-modern disposition, than it is a sign that they recognise the benefits of navigating in ways that make them eligible for help. In Delhi in the 2000s, however, the space to navigate strategically became increasingly smaller for people living in paralegal colonies such as slums and jhuggi jhopris.

The population of the National Capital Territory (NCT) of Delhi is about 16 million, but in terms of infrastructure the city really covers the much larger area of the National Capital Region (NCR) which has a population of about 22 million. The number of people living in slums in the NCT is widely contested due partly to disagreements about how to count the de facto population of the city but also about what constitutes a slum. A survey from 2001 made by the Delhi Department of Planning lists eight different categories of settlements, where seven are non-standard (Delhi Department of Planning 2001).

These include '*Jhuggi jhopri* clusters' (15%) and 'Slum Designated Areas' (19%), which both conform to the UN's definition of slums. 'Jhuggi jhopris' are clusters of makeshift shacks or tents typically erected in the interstices of the city along train tracks, river banks, parks and empty lots, whereas 'Slum Designated Areas' are located in comparatively upscale neighbourhoods with more infrastructure and better houses but that are vastly overpopulated and often without basic amenities. 'Rural' and 'urban villages' (11%) are villages in the NCT, some of which have been swallowed up by the expanding city. Displaced people from demolished jhuggi jhopris, slums or villages are resettled in 'Resettlement colonies' (13%). At the other end of the spectrum of settlements we have the 'Unauthorised Colonies' (5%), which have been built illegally but are too upscale to be deemed 'jhuggis', 'slums' or 'villages' (rural or urban). 'Regularised Unauthorised Colonies' (13%) are illegally built settlements that have been legalised retrospectively, and this leaves only 24% to live in settlements described as 'Planned Colonies'. This means that between 34% and 58% lived in slums or slum-like conditions in 2001 in the NCT.

Ten years later, Delhi is a different city. A report published 2010 on 'Urban Slums in Delhi' (Ali 2010), sets the percentage of inhabitants living in slums in Delhi NCT at only 19% and this figure is startling, given that Delhi experienced a population increase of approximately 50% (an increase of 5 million inhabitants estimated from the 2011 census). 'Planned colonies' in the NCT were not built on a scale to accommodate anywhere near that number of people, so where did they go? The answer can be found in a technicality in the 2010 report: it only counts inhabitants of slum *clusters* of at least 20 households, and this excludes a very large number of inhabitants: the people living under flyovers, in train stations or on the street, as well as the people who have been forced to move to areas on the fringe of the city outside the NCT, like Gurgaon or Faridabad. This is important because slum clearance increased from 51,461 houses demolished in the 13-year period between 1990 and 2003, to more than 45,000 demolished homes in the next 5-year period between 2003 and 2007 (Batabyal 2010, 106), leading to an all-time crescendo around the time of the preparations for the Commonwealth Games held in Delhi in 2010. Based on the increase in population, the rate of slum demolitions and the low estimate of slum colonies in 2010, it seems an inevitable conclusion that the number of people living in slum-like conditions in the NCR in 2010 was much higher than in 2001, even though the number of official slum *areas* in the NCT had decreased.

The judicial processes that legalised these demolitions were very often Public Interest Litigations filed by Resident Welfare Associations and 'leagues of concerned citizens', with comparatively privileged access to courts, due to their education and material wealth (Chakrabarti 2008, 101; Baviskar 2003, 90; Sundaram 2009, 71). Adjust Rehabilitation Policies were put in place to secure former inhabitants of slums with a legal alternative to squatting, after they had been evicted. When rehousing was provided, it was almost always located at the margin of the city, but in many cases rehousing wasn't provided because courts agreed that squatters were criminals grabbing land, rather than destitute, and thus superseded these policies. Framing this within Chatterjee's conception of political society, it is exactly policies like the Adjust Rehabilitation Policies that give inhabitants of slums an incentive to cast themselves in the role of a 'population' that must be helped by this special provision. An exception is made to provide them with a legal alternative to their illegal squatting, but this is being countered by a judicial system that overrides these exceptions by pointing to the laws of private property that are the cornerstone of civic society, giving Resident Welfare Associations the right to evict squatters from the property they own.

But this division of the courts acting in the interest of 'the public' (defined as the middle and upper class) by employing rationales of civic society, while inhabitants of slums employ those of political society, is far from consistent. The Delhi Department of Planning Report (2001) tells us that Delhi in 2001 consisted of 13% 'Regularised Unauthorised Colonies', which is to say illegally built settlements that were legalised retrospectively, and this means that the courts are not averse to employing the politics of exceptions either, when they believe it works in the interest of the middle- and upper-class 'public'. Studying how mega-malls on prime property in Delhi are routinely legalised after they have in fact been built, D. Asher Ghertner summarises the courts' use of exceptions, when deciding whether an illegal structure should be regularised in this way: 'if a development project looks "world-class", then it is most often declared planned; if a settlement looks polluting, it is sanctioned as unplanned and illegal' (2011, 2).

For Delhi residents living on a property whose legality is disputed in this era, it is thus quite important that their property is discursively framed as 'world class' or at least not the opposite, whereas members of Resident Welfare Associations would have a vested interest in casting what they perceived as inhabitants of illegal slums on their property as problems to be solved rather than citizens to be protected. As Chatterjee reminds us, the classic way to do this is by deploying strategic conceptions of modernity and defining something as its opposite, and within this general discourse, Amita Baviskar (2003, 90) identifies what she calls 'bourgeois environmentalism'. It is a discourse that fights to clean up what is perceived as pre-modern 'clutter' in the city, using the tropes and metaphors from other kinds of environmentalisms to argue for solutions that are not necessarily good for the environment and certainly not for those who are perceived as 'clutter'. Consequently, large factories polluting the Jamuna River might be perceived as 'modern' and thus 'non-polluting' within this discourse, while small-scale industries with a very low carbon footprint situated in jhuggi jhopris, slums and rural/urban villages might be perceived as pre-modern and thus 'polluting'.

This in turn means that tourists with the urge to reach across to valued but poor 'others' or 'sames' through the framework of globalised slum tourism can never be sure whether their actions, in fact, help them or not, as even an unequivocal show of public solidarity might be taken as proof of the slum inhabitants' illegality, even if this is not the intention. Consequently, this burden also befalls the NGOs facilitating this kind of tourism, as well as the journalists publicising it and the researchers of slum tourism trying to conceptually frame the experience and its repercussions.

The next section explores how a specific NGO conducting tours in a Delhi slum that ended up being demolished navigated this discursive minefield.

Figure 1.2 Demolition of Akanksha Colony – salvaging water storage utensils, captured by Jessie Hodges

Figure 1.3 Demolition of Akanksha Colony – salvaging building materials and utensils, captured by Jessie Hodges

Figure 1.4 Demolition of Akanksha Colony – police and bulldozer, captured by Jessie Hodges

Figure 1.5 Demolition of Akanksha Colony – salvaging doors, beds and LPG-canisters (gas), captured by Jessie Hodges

Salaam Baalak Trust's City Walk and the demolition of the Akanksha Colony

On 17 March 2010, a small colony of jhuggi jhopris (slums) within New Delhi Railway Station called the Akanksha Colony was demolished. Bulldozers knocked down the sheds raised along the tracks and after having picked their belongings out of the rubble, the now former inhabitants grudgingly moved east to the margin of the city, where provisional accommodation had been set up for them.

In many ways, the Akanksha Colony is the textbook example of the life and death of a slum. Thirty years prior to its demolition it was settled by railway workers who cleaned the coaches that rolled in on tracks laid right beside their small concrete huts, which lacked sanitation and only had the electricity they could illegally tap from nearby sources (Hodges 2010, 108). Their labour did not earn them enough pay for accommodation anywhere near their place of work because real estate prices were too high around the New Delhi Railway Station. Instead, they built Akanksha on a piece of land belonging to the Indian Railway Catering and Tourism Corporation (forthwith IRCTC) and because the IRCTC needed the services provided by the labourers, they kept the colony in a state of permanent paralegality, never quite legal, never forcibly removed.

In this way Chatterjee's description of the paralegality of squatting rings true for the inhabitants of the Akanksha Colony:

> There are obvious reasons why population groups belonging to the urban poor could not be treated at par with proper citizens. If squatters were given any kind of legitimacy by government authorities in their illegal occupation of private or public lands, then the entire structure of legally held property would be threatened.
>
> (Chatterjee 2004, 136–37)

Cast within the conceptual framework of Chatterjee, the inhabitants of the colony turned from the language of bureaucracy and legislation (referring to their civic rights as citizens) to the language of politics and a 'moral rhetoric' (referring to the rights they *ought* to have) (2004, 60), thereby entering into 'political society'. In both the writings of Chatterjee (2004) and Baviskar (2011, chap. 15) ethnographic studies show that influence is sought by inhabitants of paralegal colonies, whose ways of life might otherwise be viewed as too pre-modern for a 'world-class city', by reaching out to local politicians, who might be induced to make an exception and spare the colony from 'development' and bulldozers. The inhabitants of the Akanksha Colony attempted to get their colony legalised by negotiating with the IRCTC as unionised railway workers, but this was a difficult position to maintain as many of the original inhabitants' descendants living in the colony in 2010 had little or no connection to the IRCTC. Consequently, they lost, and the colony was demolished.

The demolition wasn't exceptional but, rather, an all-too-common part of the hard everyday life of Delhi's poorest inhabitants. One might even say that it really only differed significantly from all the other demolitions in Delhi at that time in that it altered the only slum tour in Delhi; the 'City Walk' (CW) operated by the NGO, Salaam Baalak Trust (SBT). Before the demolition, the route of the CW started outside the Railway Reservation Centre on Chelmsford Road and proceeded into the Akanksha Colony. Here, SBT had set up a contact point with a day school, where the children from the colony could receive education and CW-visitors, who were almost all tourists from the global North, were invited to interact with whoever was around at the time while being told about the area and the work the NGO was doing there. The CW would then proceed to the part of the station where passengers generally went, and here the guides would tell about their own lives before they joined SBT. To illustrate this, they would interact with current bands of street children who survived on the station by collecting trash and selling it to trash dealers in nearby Pahar Ganj. They would point out where they used to sleep in the hollow in between two overlapping roofs where the footbridge met the platform. The hollow was so small that adult-sized people couldn't enter, and this fact would often be used as a stepping stone towards narrating about incidents from their lives that might help the visitors understand the precariousness of street life for children living without any safety net. An example of this could be stories of how they were robbed of their belongings while sleeping at the station and how this established a sense in them that it was not only futile but also downright dangerous to accumulate even small amounts of material wealth and that this, in turn, would give rise to a belief that planning and looking ahead was useless, a belief that made it hard for social workers to ask them to do precisely that by joining various NGOs, of which SBT was one.

Meanwhile, current street children living at the station – some as young as six years old – would usually come up to the guides with plastic bags filled with correction fluid covering their noses, removing them only long enough to exchange a bit of banter and ask the CW-visitors half-heartedly for some change in Hindi. They were routinely sent away by the guides with the instruction that they could join the organisation and go into rehabilitation for free but that they wouldn't get any money. As poor as the children living in the Akanksha Colony might be they were still far better off than the children who lived on the station without the social structure of a family, even if the children in Akanksha Colony lived in housing one could easily call 'slums'. As previously argued, a major part of Delhi's population lives outside or on the margin of the 'official city' and without glamorising it, the CW thus tried in a small way to show how what might be called 'street life' or 'slum' is really a set of living conditions as diverse as those found within the 'official city'.

Leaving the platforms, the CW would proceed to an SBT Contact Point on the first floor of the General Railway Police office just outside the station building, where medical facilities, day school, food and entertainment were provided for street children in the daytime. After a bit of interaction with the staff and children, the group would then be led into the *galis* (alleys) of Pahar Ganj, past some *dhabas* (cheap eateries) where current street children might find work, a 'trashroom' where they might sell whatever they collected at the station and, finally, to the *Aasra Centre*, which is a somewhat rundown shelter home that still serves as the headquarter of SBT. Here, the CW-visitors would be encouraged to interact for 15 minutes with the fifty-odd boys staying at the shelter home, and this, along with the payment of the 200 rupee ($5) fee at the accounts office, would conclude the CW.

With the demolition of the Akanksha Colony and the restriction of access to the Railway Station, the CW had to find another route while trying to show some of the same activities and explain the same dynamics. The new route still had the railway station as a pivotal site, and similarly it still dramatised the progress of children, who ended up in SBT, but it had very few illustrations of the life they led before. Like a commercial featuring a 'before-and-after', it now missed its 'before' as CW-visitors only interacted with children who had already been 'saved' by SBT. The bright young men standing in front of the CW-visitors delivering the script now revealed little of the journey some of them had made from drug rehabilitation, psychiatric evaluation and treatment and the painfully slow re-socialisation into an environment where they might feel so safe that they could eventually start telling a personal story that was not entirely made up to support whatever position they wanted to inhabit in relation to the NGO. Likewise, the important difference between 'street life' within the social structure of a colony like Akanksha versus living alone in the station was hard to illustrate without showing life in the Akanksha Colony.

The demolition meant not only a shift in route for the CW but also a shift in their representational praxis, which also shifted the discussions of moral and ethical issues connected to this. In an interview conducted 3 years after the demolition, the CW- and volunteer-coordinator from 2009–2010, Jessie Hodges formulated it in this way:

> when we were going inside [Akanksha and the station] talking about street life it was borderline voyeuristic and going around these slums, you know Akansha, and inside and pointing to street kids and saying, *'That's what they're doing, look!, look there now!, he's* you know *doing drugs right now and* look *here! and, like, that kid is . . .* ' I mean it was borderline, so that was like a different show.
>
> But then, I mean, when you couldn't go inside the street or railway station then suddenly, like, you wanna tell people about street life but you're in a market place, you know, you have no example in front of your face and the people have no context within which to put your stories, so then naturally your stories change to fit your context a little bit more.
>
> (Interview with Jessie Hodges)

Before the demolition, SBT had been faced with charges of voyeurism from journalists (see for instance Gentleman 2006) and because of this, SBT-staff and guides frequently discussed the ethics of showing CW-visitors 'street-life'. An elaborate set of instructions was in place to make sure that the interaction between street/SBT-kids and tourists/CW-visitors was conducted in a way where neither party felt exploited nor exploit*ing*, but even so it was a concern among the staff, as we might see from Hodges's admission that it was 'borderline voyeuristic'. As has frequently been documented in studies of slum tours and the debates they give rise to (Selinger and Outterson 2011; Ma 2010), one argument against slum tours is that the locals are not in a position to refuse participating in the encounter, making the question of whether they give consent to do so meaningless. In the criticism levelled against SBT and the CW, it was less the locals of Pahar Ganj that were thought to suffer from the perceived voyeurism of the tourists than the children living on the street or at the SBT shelter homes, and it was thus argued that it was the responsibility of SBT to shield them from the gaze of the tourists, since they were in no position to choose for themselves, being both minors and at the mercy of SBT.

According to Hodges, the demolition changed this and suddenly posed the opposite question: Do you have an obligation to show 'slums' and 'street-life' in all its horror, beauty and everydayness?

If it had been the case that the slum simply disappeared and better accommodations were systematically provided to its former inhabitants, the question might have been easily resolved; SBT could go back to its main purpose, caring for street children, and stop conducting tours into slums that were no longer there.

But as we have seen that was not the case; rather, it seems that the city was trying to displace its poverty-stricken population to its margin rather than provide them with real solutions to their problems, and showing just what grave consequences this had for the displaced population of former slums was thus more important than ever. But though the widespread demolitions of slums highlighted the necessity to represent them, it also proved to be one of the biggest obstacles to doing precisely that.

First, there were no guides to represent the demolished slum on the CW. The former inhabitants of the Akanksha Colony were resettled on the eastern rim of the city. To the majority of the inhabitants however, this meant that they either had to give up the jobs they previously held, as the commute to New Delhi was too long and the fee too large an expense to incur on their salary, or find other types of informal accommodation closer to their jobs. Many chose the latter, but to the only two CW-guides who were actually living at Akanksha this was not an option as they had to stay with their families, and the only place they could do that was where they had been relocated to. The CW thus lost not only the opportunity to show CW-visitors the difference between life at the Akanksha Colony and the Railway Station; it also lost the only two CW-guides that might give a first-hand account of what it was like living at Akanksha, as well as what it had meant to them that the colony had been demolished.

Second, setting up the CW in the Akanksha Colony responsibly had taken a long time and a new slum where SBT had the same contacts could not readily be found. SBT has as a specified goal that its tours should provide contact with a given slum's inhabitants through a framework that also worked towards the betterment of their living conditions and showed them as resourceful members of society. Most slum-tour operators have goals like these, which are publicised to a varying degree. One example is given by Meschkank (2012) study of how 'Reality Tours' in Mumbai semantically produces *Dharavi* as a place marked by myriads of proliferating small-scale industries, in order to counteract the stereotype of a slum as a place of stagnancy and inactivity. It could, however, be argued that SBT went even further to establish the CW as a space of cultural interaction marked by mutual respect. SBT worked with street children for 19 years in Delhi before showcasing its work to tourists via the CW in 2006, and today only a handful of SBT's 150 employees work with the CW at all while the rest run SBT's five shelter homes, 12 contact points, mobile schools and other social initiatives. SBT is thus careful to present the CW as a spin-off from social work, and not the other way around, and SBT's high standards for the kind of social work they would be willing to showcase in a slum as well as the degree of understanding they felt they would need to have with its inhabitants meant that they in 2014 still haven't found a suitable replacement for the Akanksha Colony as topos marked by 'slum' that can be showcased on the CW.

Third, the need to showcase certain slums as viable alternatives to other types of urban development was also countered by the fact that a showcased slum would be in danger of demolition because it was being showcased. It was more than likely that a new CW-route going through a slum area would draw the Delhi

Development Authority's (DDA) attention to the fact that its inhabitants were squatting illegally and that this, in turn, might provoke the DDA to order the slum demolished, as it wouldn't fit the image of a 'World-Class City'.

Conclusion

The chapter shows that slum-tour operators face the epistemological questions raised within subaltern studies concerning the limits of ethical knowledge production about slums and their inhabitants. In the concrete case of SBT's involvement with the Akanksha Colony, the two moral obligations, to *show* or to *shield* from view, are two opposing ethical demands that coexist within SBT, along with extensive discussions of what, how and why this should be done.

As the debate over voyeurism represented earlier shows, main actors represented within the show/shield debate are most often (1) the slum's inhabitants, (2) the slum tour operators and (3) tourists from the global North. But the demolition of Akanksha, the repercussions it has for the CW and the geopolitical climate it takes place within, highlight that what is shown (or shielded from view) to tourists from the global North is also shown to the governing authorities of the city that the slum tours take place within, especially if those slum tours are also represented and promoted via the media or discussed within slum tourism research. The epistemological question posed previously is thereby not just a question of how organisations like SBT might produce knowledge about slums and what that knowledge might be used for. It is a question that might also productively be posed to the entire field of slum tourism research: How do we produce knowledge about slums that does no damage to the people that live in them or the organisations that try to faithfully represent them?

The next chapter shows that SBT decided not to continue the practice of the CW in another slum, or to continue representing the demolished slum they formerly operated within, because they concluded that the ethical demand to shield from view was greater than the demand to show. The route of the CW thereby changed so that it wound through another topos, and the CW's focal point changed to how the former street children acting as guides represented this new topos and SBT as an organisation, by referring to their former identities as street children. Far from rendering the show/shield debate moot, however, the shift in topos shifted the focus of the debate to centre on the moral implications of how the guides co-performed their identities with the CW-visitors and how this positioned guides, visitors, current street children and SBT within touristic and humanitarian logics.

Notes

1 For an overview of Said's original theoretical approach see *Orientalism* (1979), pp. 1–28.
2 In its earliest form outlined in *Culture and Imperialism* (Said 1993) this theoretical approach to reading fiction from the colonial centre actually bears much resemblance to the Subaltern Studies' approach to reading documents from colonial archives, in that they both focus on the subtle signs of the coercive relationship between coloniser and

colonised that might be invisible for anyone not looking for them, but which might expand the reader's understanding of them if they are included in a contrapuntal perspective.

3 He elaborates this argument most consistently in *Lineages of Political Society* (2013a), taking an outset in the debate over the conduct of the governor general in India in the 1780s, who insisted that India could not be governed according to the laws of England, especially if Britain wanted the revenues from its overseas possessions to flow as freely to the metropolis as they did at the time. This critique inscribes itself in to a tradition of critiques of the origins of universalist principles at the heart of modernity, from Edward Said's critique of the general imperialist attitude to an 'orient' it constructed in order to dominate it (1978, 1993) to Dipesh Chakrabarty's (2000) critique of John Stuarts Mills's Eurocentrism when he draws up the principles of liberty and democracy while arguing that it could not yet be extended to colonies like India. To this list could be also added David Harvey's critique of Kant's universal ethics and cosmopolitan principles, which either work towards eradicating cultural and geographical difference or 'operate as an intensely discriminatory code masquerading as the universal good' (Harvey 2001, 112), as well as David Theo Goldberg's critique of Locke's famous *First Treatise of Government*, where he argues that slavery cannot be justified, except if the slaves are taken in 'just wars' or if the slaves are not capable of reasoning. Goldberg concludes that 'the concept of race has served, and silently continues to serve, as a boundary constraint on the applicability of moral principle' (1993, 28). Referring to the case at hand, we might conclude that the boundary constraint on the applicability of moral principle within the Indian metropolis extends to a number of 'othered' subject positions which are not typically race-based, though they might very well include the question of caste.

4 It is important to note that both Chatterjee and Anderson analyses the nation state as imagined – which is to say envisioned – and when Anderson writes of the nation's 'deep, horizontal comradeship' (1991, 7) he sees it as something that paradoxically overshadows the 'actual inequality and exploitation that may prevail' (Ibid.) in the minds of the nation's inhabitants. Chatterjee's point is thus not simply that the homogeneous conception of the nation state can act as an instrument of oppression, as Anderson acknowledges this, but, rather, that subjects within most nations in the formerly colonised world *only* define themselves as belonging to this homogeneous space/time of the nation state (as e.g. 'citizens') when they are allowed to do so and when it suits them.

5 B.R. Ambedkar, Independent India's first law minister and the principal architect of the Constitution of India, who was himself a Dalit.

6 These might be based on the community they have built within the colony, a shared religious affiliation, a political party they might support or a shared ethnic background in that they for example are displaced migrants from Bangladesh.

2 The authentic slum or former street children as prisms of authenticity?

Conceptualising Delhi's informal urbanism as a creative, subaltern space

Since Chatterjee wrote the *Politics of the Governed* a number of studies that draw implicitly or explicitly on his theoretical framework have emerged with Delhi as their empirical focus. Baviskar (2003) links the issue of slum demolitions in Delhi to a long-standing practise of not including working class habitation in Delhi's city planning, thus forcing the working class to dwell in the interstitial spaces of the city, and she traces it back to the planning and construction of Edwin Lutyens's New Delhi (1911–1930), whose colonial, large-scale architecture aimed to get rid of the spatial confusion of Shahjahanabad (now Old Delhi), where residential, industrial and trade zones then as now intermingle in a typical bazaar structure. The poor lower-caste population, who lived where New Delhi was supposed to lie, was summarily moved to the land west of the new city (2003, 91), and Baviskar reads this as a classic example of how city planning often doesn't rid the city of slums and illegal construction but, rather, *produces* it, because the people who are displaced always need to reside somewhere and will thus typically settle at the margins of the city in slums.[1] In 1911 this was furthermore exacerbated by the fact that Delhi actually needed workers to build and maintain the new city, and as there was no place for them in its plan, they were forced to live in temporary dwellings near the building sites.

Drawing on Chatterjee and the Subaltern Studies approach,[2] Ravi Sundaram (2009) shows how this practice continued up through the 20th century, though the avowed reasons for doing so in independent India had more to do with the wish to create socially viable urban environments than the imperial spectacle of New Delhi erected by the Raj. Social reformers such as Prime Minister Jawaharlal Nehru and novelist Mulk Raj Anand[3] were deeply committed to creating a modernist city,[4] which nonetheless instituted a segregation of space that he calls 'urban apartheid' (2009, 55), as large sections of people too 'rural' in their way of life were moved to the outskirts of the city in specially created 'urban villages' – often without their consent. Sundaram's and Baviskar's assessment that the city needs the working-class population it marginalises is backed up by Kaveri Gill's (2012) study of Delhi's trash collectors, as well as studies that focus on the environmental consequences of building a car-friendly city that welcomes big industrial plants along the Yamuna River while closing smaller workshops in residential areas, which might have had health and safety issues but also a much smaller carbon footprint (Chaturvedi 2010).

Crucial to the topic of this book, these texts seem to be part of a new discourse within studies of informal urbanism that sees the heterogeneous time/space of the Indian nation as holding a creative potential. Baviskar (2011) shows that the cows of Kamla Nagar in North Delhi, who wander the streets among the more structured forms of traffic, are not reined in because some of the residents of Kamla Nagar act as strongmen to a political party, but she simultaneously conceptualises the cows as an Indian version of de Certeau's flaneur (Certeau 1984, 92). So while she recognises the patron–client relationship that Chatterjee theorises as a way for subaltern groups to gain influence, she also uses the informality of livestock in the city to articulate a critique of the idea that the vision of the transparent, panoptic city can ever be achieved, even if it is created by thinkers like Nehru and Anand, who see it as a means to combat social inequality.[5]

Similarly, Sundaram concludes that cultures of piracy in Delhi's informal sectors work as a means of survival for subaltern populations (Sundaram 2009, 19–21), but he also sees them as a means of self-expression that rewrites notions of the homogeneous space/time of the modern the nation state. In Arjun Appadurai's idea of 'The Social Life of Things' (Appadurai 1988), objects have a 'life-cycle' where they become commodified, purchased, used and, after a time, thrown out and thus turn from 'things' into 'waste'. Sundaram points out that within this logic, things made from recycled waste might thus be seen as 'reincarnated', but he furthermore expands this vision of the 'world of things' as he writes,

> The world of piracy ranges from not just immaterial media goods of all kinds (software, movies, music, hardware) but also most mass-market commodities ranging from the counterfeit to the 'unbranded', the 'graymarket' or the local commodity.
>
> (Sundaram 2009, 12)

This extended conceptualisation of piracy, where different shades of production and reproduction bleed into each other, challenges the idea of the 'copy' and thereby the multinational, brand-based economies that live off the notion that they can sell identical copies of commodities all over the world. The *'pirate modernity'* ensuing from this type of informal urbanism and illicit modes of production may therefore 'be seen as globalization's illicit and unacknowledged expression' (Sundaram 2009, 14). This thereby also shifts perspectives of the agency possessed by inhabitants of informal urban areas and workers within the informal sector, and this has led to discussions about how to theorise them.

But if certain academic studies have cautiously located a creative potential within informal urbanism, then popularised media forms have gone even further and represent the Indian slum and interstitial spaces of informal urbanism as topoi positively imbued with excitement, and while it would be impossible within the scope of this text to mention them all, this small, incomplete overview of texts might give a sense of the proliferation of the positive approach to informal urbanism. In an Indian context, the place most often represented is Mumbai, and especially the large slum Dharavi. Academic studies of the area include Sharma's *Rediscovering Dharavi: Stories from Asia's Largest Slum* (2000) and Verma's more generalised *Slumming India: A Chronicle of Slums and their Saviours* (2002), which soberly try to engage with the problems faced by inhabitants of slums.[6] Recent years, however, have seen the emergence of sassy coffee-table books such as *Poor Little Rich Slum* (Bansal 2012), where the endurance and ingeniousness of Dharavi's inhabitants are represented in colourful photographs that emphasise the image of the wealth-generating slum. Because it attempts to remain apolitical in its scope, it inadvertently taps into a neo-liberal discourse of the slum dwellers' brave struggle to succeed as individuals and thereby a scope first popularised by *The Economist* (The Economist 2005), where the acknowledgement that creativity and productivity arise in response to hardship does not result in a critique of the global economic system that makes precarious work conditions preferable to no work at all. Most famous however has been the Oscar-winning motion picture *Slumdog Millionaire* (Swarup 2009), whose impact on the tourist industry has been argued by Sengupta (2010), or the novel *Shantaram* (Roberts 2003), which is also connected to the tourism industry by Meschkank (2011, 52; though Ma (2010) disputes this) or, indeed, Kathrine Boo's later fieldwork-based novel *Behind the Beautiful Forevers* (2014), where the fusion between the slum as an object of academic inquiry and a dangerous/desirous space of exploration for the foreigner has been naturalised completely.

The informal urbanism of Delhi has generally generated fewer publications and less attention, though one should perhaps mention the Nobel Prize–winning novel *The White Tiger* (Adiga 2008), the collection of essays called *Trickster City: Writings from the Belly of the Metropolis* (Sarda 2010) or the foreigners' view of the city in *Delhi: Adventures in a Megacity* (Miller 2010), which reads like an updated version of Dalrymple's classic travel account, *City of Djinns: A Year in Delhi* (Dalrymple 1993). As argued earlier, Delhi has had a long history of demolishing large slum areas, and this has led to a significant difference between Delhi and Mumbai, in that Delhi's informal sector is located in the city's interstitial spaces or at its margins, whereas vast areas of informal habitation such as Dharavi have become landmarks that characterise Mumbai to foreigners in influential media, the hidden dangerous/desirous *Belly of the Metropolis* (Sarda 2010) is located in Delhi's interstitial spaces. Consequently, Delhi's slum or poverty tourism – or perhaps more widely tourism aimed at exploring informal urbanism[7] – locates its exploration not in vast slums but in the back alleys of the official city, where instances of the creatively heterogeneous space/time of the metropolis is mediated to tourists, either as obvious spectacles or as cultural practices that are unintelligible to tourists if they are not accompanied by guides who are 'in the know'.

Ravi Sundaram (2009) characterises this overall discourse of marginalisation as instances of 'urban crisis writing' that institutionalises urban fear of the informal, and he sees this tendency extended primarily in the Indian popular media.

However, academic surveys of slums such as Mike Davis's *Planet of Slums* (2006) are also complicit in Sundaram's view, because he believes Davis's discourse heralds 'a new apocalyptic "slumming" of the world's cities' which will inevitably lead to fear of the 'slum' and consequently the extreme marginalising measures described by Chatterjee and Baviskar earlier. One of Mike Davis's central arguments is that surplus population generated by rural–urban migration in the global South will result in the emergence of slums around cities where a sufficient degree of industrialisation has not yet taken place. The problem is thus not only the lack of plumbing and planning of residential areas leads to a lack of sanitation and high-rises but also the lack of factories leads to lack of employment within the formal economy. China is highlighted by Davis's as the only place where Marx's proletariat exists today, because it is the only nation that has had the foresight and means to build not only factories and high-rises to the workers migrating from rural areas but also an infrastructure to distribute the goods produced there to the world.

To Sundaram, this diagnosis seems to necessitate the continuation of the planning and building of the postcolonial, panoptic city, which in the case of Delhi transforms long-standing slum areas into the temporary *juggies* of otherwise homeless people. What Davis argues is that celebrating 'informal urbanism' in the slums does not better the actual living conditions there but only amounts to an exotisation that shirks the responsibilities states have to do so. He quotes Gita Verma's book *Slumming India* on Delhi: 'The right to stay is no great privilege ... it may stop the occasional bulldozer but, for the rest, it does little beyond change the label from 'problem' to 'solution' (Davis 2006, 78). It should be noted that the discourse Verma is speaking against is not so much Sundaram's perspective but, rather, the type of writing that celebrates the 'money-minting slum' in mainstream media (The Economist 2007) or *jugaad* (making do) as a business model that will in itself create greater equality (see also the discussion between Birtchnell 2011 and Radjou et al. 2012). Sundaram does not defend the of lack social awareness that this type of neo-liberal writing displays but, rather, argues that chronicling the horrors of the slum will not in itself ensure that future development plans will alleviate it, as past plans have proved to do the opposite, though the intentions behind them may have been noble enough.

What we might preliminarily conclude is that most knowledge produced about a given 'slum' in Delhi in the 2000s strategically deploys discourses regarding its modernity, orderliness and even its environmental sustainability because all these parameters might become factors in the evaluation of whether to demolish that slum or not. If the discussion between Davis/Verma and Sundaram is framed by Chatterjee's division between citizens and populations (see Chapter 1), it also means that the production of knowledge about slums might be critiqued not only for its actual content or political intentions but also for what subsequent actions it might be used to justify later. Simply asking the people living in poor, urban areas whether they think they are living in a 'slum' is not a fail-safe solution either because they strategically navigate the elite discourses of citizen/population according to the benefits they might gain in a given situation.

Discursive and performative approaches to studying tourism

John Urry conceptualises 'being a tourist' as employing a 'tourist gaze' (Urry 1990), through which a topos and its inhabitants are perceived. This gaze is disseminated via a cycle of representation, where tourists travel to places mediated to them as tourist destinations, snap photos of them and share them, so that other prospective tourists might look at them and repeat the representational cycle. The tourist gaze, as Urry defines it, is thereby not determined by the individual but is a social construct shaped by numerous representations and re-representations of places that gradually turn into tourist destinations, when subjects situated as tourists interact with them as such. Conceptually, the idea of the tourist gaze draws on Foucault's study of heteronormativity, where he examines its pathologising gaze on homosexuality (Foucault 1978), as well as his study of forms of state control, where he examines the ordering and penalising gaze on the criminal (Foucault 1977). In a similar vein, Urry attempts to study the 'everyday' by studying what is defined as its opposite: the 'holiday'.

After the first version of *The Tourist Gaze* was published, its methodological scope was critiqued for not incorporating a performative approach (e.g. Perkins and Thorns 2001), as it tries to conceptually grasp how subjects discursively positioned as tourists actually perform in the field. In the subsequent rewritten version (Urry and Larsen 2011) the authors attempt to mediate between the two and write,

> The Tourist Gaze 3.0 rethinks the concept of the tourist gaze as performative, embodied practices, highlighting how each gaze depends upon practices and material relations as upon discourses and signs.
>
> (Urry and Larsen 2011, 15)

True to the title of the book, however, the analyses of tourist performances presented in *The Tourist Gaze 3.0* seem to centre on visual senses, rather than the gustatory, olfactory, auditive or tactile senses, though these are mentioned as important. Similarly, as the quote indicates, the inclusion of tourist performances as an analytical category seems to work primarily as an illustration of how they affect the tourist gazes as an overarching category, and the book therefore focuses less on how to approach studies of concrete tourist performances methodologically and more on how these performances constitute gazes.

The performative tradition Urry and Larsen attempt to incorporate into their theoretical approach is partly founded on Goffman's symbolic interactionism (1959) that uses the vocabulary of theatre performances to describe everyday transactions. Timothy Edensor (1998, 2001, 60) uses Goffman's framework but employs a different vocabulary, as the subjects Goffman analyses seem to play a set of roles while their 'true self' is positioned safely somewhere behind these, while Edensor employs a framework where the individual seems to *consist* of a series of subject positions available to it at a given time. This is consistent with how Foucault generally theorises the identification process and Edensor's perspective thereby translates relatively easily into Urry and Larsen's reconceptualisation of the tourist gaze as performative, embodied practices (Urry and Larsen 2011, 198–216). Drawing on a similar tradition, Edward Bruner views 'performance as constitutive' (Bruner 2004, 5, 28) of the identification process and he agrees with Urry's claim that subjects are increasingly invited to employ a tourist gaze on what would otherwise be their everyday, but this does not mean that the only performances subjects can participate in are touristic performances. Framing the guided tour as a performance Jane Meged (2010) draws on Bruner and Edensor to show that each tourist performance is *co-produced* by its participants, who are allowed a measure of situated agency to create it anew each time, though this is done in relation to a series of dominant discourses.[8]

Combining Meged's perspective on the guided tour with Chatterjee's perspective on the situated agency of contemporary, urban subalterns from Chapter 1, I conceptualise the co-performed, guided slum tour as one of the arenas within which urban subalterns might articulate their allegiance to certain subject positions within discourses framed by either political or civil society. The next section of this chapter is partly devoted to an analysis of how guides situated as liminal urban subalterns do this in practice, while Chapters 5 and 6 focus on a similar dynamic in the case of tourists.

Returning to a conceptual framing of tourism studies, the epistemological framing of tourism as a performative practice also has a direct bearing on how spaces and places might be conceptualised as tourist sites. Since mass tourism became popular in the global North in the 1960s, two parallel developments seem to have taken place. On one hand, increasingly 'independent' kinds of travel and tourism have been developed, marketed and sold as alternatives to the kind of tourism that Daniel J. Boorstin's famously called 'Pseudo Events' (1662). At the same time, academic analyses of tourism have struggled to find ways of ascribing some sense of perceived authenticity to the experiences tourists have – commodified and performed though they might be – while simultaneously critiquing the idea that each new independent travel form is somehow more 'real' than all the previous ones. A landmark in this respect is Dean MacCannell's oxymoronic notion of 'staged authenticity' (MacCannell 1976). He sees Boorstin's critique of tourism as elitist, because it implies that the tourists that seek 'Pseudo Events' are then always the Other, inauthentic people, and he writes,

> Sightseers are motivated by a desire to see life as it is really lived, even to get in with the natives, and at the same time, they are deprecated for always failing to achieve these goals. The term 'tourist' is increasingly used as a derisive label for someone who seems content with his obviously inauthentic experiences.
>
> (MacCannell 1976, 92)

MacCannell is not derisive of the tourist's urge for authenticity, but drawing on Goffman's front- stage–backstage metaphor[9] he concludes that the touristy quest for a 'backstage' most often results in the staging of this backstage, which can then be presented as hitherto 'undiscovered' and thereby 'un-touristy'. But the continual discovery of this backstage by successive groups of tourists results in its staging becoming increasingly ritualised so that only certain aspects of the everyday life of the other is made visible from specially made, cordoned-off vantage points.

In contrast to MacCannell, Edward Bruner dismisses the concept of authenticity completely, partly because he sees it as perpetuating colonialist discourses of the pre-modern, authentic, exotic other, and he therefore wants to distance himself from a tourism industry that pretends to be visiting 'replicas of life in the ethnographic present, static, timeless, without history, without agency, without context' (2004, 4). His examples are from Indonesia, but Bhattacharyya (1997) sees the same dynamics in an Indian context, where tour operators beckon customers with tours to a timeless Hindu past, which bears a clear resemblance to the representational practices identified in Edward Said's classic study of *Orientalism* (Said 1978). Bruner, however, also wants to distance himself from Boorstin's view that these representational practices are 'Pseudo Events' or Baudrillard's view of them as *simulacras*, because he believes that these ideas presuppose that the 'original' space exists somewhere but that tourists simply don't have access to it. Instead, he believes that the 'basic metaphor for tourism is theater' (Bruner 2004, 209) and just as looking for absolute truth in theatre misses the point of the format, so authenticity is a 'red herring' (2004, 5) that confuses the discussion of what the tourists are experiencing, because the tourists themselves have gone beyond this unproductive discussion in his opinion, and Bruner therefore simply refrains from introducing the concept in the interviews he conducts.

To Bruner, performance is thereby constitutive not only of subject positions inhabited in the identification process taking place during the tourism performance but also of the spaces inhabited while there, and this means that Bruner also departs somewhat from Urry's definition of a holiday as the leaving of a 'home' and the return to it (Urry 1990, 1). Bruner accepts Urry's assertion that tourist destinations are socially constructed, but he questions whether tourists then really just go 'away', or whether they, in fact, visit a version of 'home' created especially for them. The tourism Bruner analyses 'occurs on a borderzone physically located in an ever-shifting strip of border on the edges of Third World destination countries' (Bruner 2004, 192), and this space is constituted by the infrastructure set up for the benefit of tourists, such as transport lines to and from it, accommodation which caters to tourists' ideas of comfort, restaurants with multilingual menus and so on. Simultaneously it is an imagined topos, where the tourists interact with versions of the destination's culture that are commodified and sold to them for a price by performers who engage in this trade. The touristic borderzone is thus

> a point of conjuncture, a behavioral field that I think of in spatial terms usually as a distinct meeting place between the tourists who come forth from

their hotels, and the local performers, the "natives", who leave their homes to engage the tourists in structured ways, in predetermined localities for defined periods of time.

(Bruner 2004, 17)

Bruner's dismissal of authenticity is somewhat problematic when analysing forms of tourism whose marketing explicitly play on constructions of authenticity. Peter Dyson productively uses authenticity as an analytical category, when analysing the company Reality Tours' 'Slum Tours' into Dharavi, Mumbai, because on this tour,

> the urban poor are framed as an unrecognized and undervalued group who live a life 'more real' than can be observed in the 'rich spaces' of the city. Indeed, the company's promotional leaflet markets the tours as a chance to 'glimpse into the reality of everyday life in Mumbai and India', with the tagline, 'See the "real" India'.
>
> (Dyson 2012, 9)

Reality Tours' marketing reads as an attempt at constructing the Indian slum as an authentic backstage to an inauthentic front stage, where the rich city shows its glitzy facades, and these constructions of the poor, informal, urban India as the 'true' or 'authentic' spaces of the city need to be critiqued, but one can only do so if authenticity – understood as a social construct that might be staged – is a part of one's theoretical framework.[10]

Fabian Frenzel's pivotal book on slumming (2016) uses Koven's (2004) and Steinbrink's (2012) studies of historical slumming to argue that the slum has always had its tourists, who helped frame it as a place apart, while the very fact that the tourists went there to experience this 'apartness' revealed it to be a fantasy of otherness, tied to a political struggle that is also described in the previous chapter of this book. Seen within this perspective, the slum can fall within Bruner's definition of a 'touristic borderzone', where the long-standing presence of tourists is continually erased by the tourists themselves, who help co-create this identity of authentic separateness from the official city. It is also true, however, that this classic staging of a backstage, performed by a continual erasure of oneself as a tourist, cannot stand alone. Frenzel provides an alternative when he analyses slum tourism through the framework of *tourist valorisation* understood broadly as a term that covers the role tourists play in turning a place into a tourist attraction, and thus, in time, a tourist commodity to be marketed and sold. Chapter 5 of this book returns to how Frenzel conceptualises the process of valorisation as taking place within the political economy of the slum and its surroundings. The point here, however, is that while tropes of staged authenticity circulate at one level, the slum might also be valorised by tourists in relation to, say, the desire for a space of leisure, marked by comfort, safety and affordability, to name but a few attributes.

The following section provides an example of how Salaam Baalak Trust's City Walk tried to cater to the opposing desires they imagined the tourists might

Prisms of authenticity? 45

Figure 2.1 Street child collecting trash in Pahar Ganj

harbour in relation to a tour of a neighbourhood that may or may not be described as 'informal'.

Street life and 'prisms of authenticity' in Pahar Ganj

Salaam Baalak Trust's (SBT's) headquarter is located in the area of Pahar Ganj, and it is the topos chosen for its City Walk (CW). The first section of this chapter concluded that Delhi's imaginative *Belly of the Metropolis* (Sarda 2010) is discursively produced as existing not in vast slum areas like Dharavi in Mumbai, but rather in the city's interstitial spaces, and the authenticity construction performed by Reality Tours and Travels mentioned in the previous section that divides the 'real' informal city from the 'fake' official city is therefore hard to maintain in Delhi. This is also the case for Pahar Ganj. The area existed as a grain market situated outside the walled city of Shahjahanabad from the 1730s (Blake 2002, 58, 117), and it therefore contains the same labyrinthine structure of alleys featured in Old Delhi (former Shahjahanabad), as opposed to the straight lines of the imperial city of New Delhi (built 1911–1930).[11] Pahar Ganj narrowly escaped being demolished to make way for the new city, and instead, New Delhi Railway Station was placed right at the entrance of Pahar Ganj Main Bazar, which has since provided the area with travellers.

Both the *Rough Guide to India* and *Lonely Planet India* have separate maps of the area drawn up in their Delhi sections, and *Lonely Planet* describes it as a place with 'bumper-to-bumper budget lodgings' and 'the place to tap into the backpacker grapevine', though 'with its seedy reputation for drugs and dodgy characters [it] isn't everyone's cup of chai' (Singh et al. 2011, 83). But, as Chris Hudson points out in his study of the backpackers' representation of Pahar Ganj on blogs, the area is not only described as a 'discovered' place, which has changed due to the influx of tourism. It is also a ' "marked" space with a reputation for the squalid that precedes arrival' (Hudson 2010, 374), and this is significant because the 'squalor' and informal urbanism described on these blogs is being discursively produced as an integral part of the culture shock that marks the beginning of the '*rite of passage journey*' for young backpackers looking for an '*anti tourism experience*' (Hudson 2010, 374).

Similarly, Kelly Davidson's study of backpackers-who-do-not-want-to-be-called-backpackers in India describes how Pahar Ganj and similar spaces in India form a backdrop for experimentations with identity for travellers:

> This delighted preoccupation with urban chaos is based on the notion that immersing oneself in India's 'otherness' and submitting the mind and body to poverty, disease and deformity will bring about a state of heightened sensory perception that crosses the boundaries of repressed Western perception. This popular outlook on the city reworks the mythologies of India as a particular symbolic space where 'beyond the boundaries' experimentations with structured 'Western' identities are effected.
>
> (Davidson 2005, 48)

So even though Pahar Ganj has been shaped by an industry catering to travellers since at least the 1930s and is thereby the very definition of a 'touristic borderzone', it might still be staged as an authentic backstage to the city but crucially not because it is untouched by tourism but, rather, because it has been specifically created to cater to tourists who want to go 'beyond' their restrictive, 'Western' identity formations by immersing themselves in an 'otherness' found within informal urbanism.

Prisms of authenticity? 47

There are, of course, also travellers, who stay in the area simply because it offers cheap accommodation, but for tourists looking for a journey 'beyond the boundaries', the street children of Pahar Ganj could be interpreted as subjects, who out of dire necessity have taken an 'authentic' version of a journey the tourists perform a codified and commodified version of, as a form of play that they can quit if they tire of it. The previous chapter described how the demolition of the Akanksha Colony and the ban from entering the railway station in 2010, erased the topos of SBT's CW, as it barred them from showing tourists how different kinds of informal urbanism, or 'street life' as SBT calls it, affected the children the non-governmental organisation (NGO) worked with in different ways. As I started preparing the main part of my fieldwork in the winter of 2013, I was interested in how SBT tried to frame the CW as an authentic encounter with informal urbanism after the demolition. Had it found a new slum to tour? As I logged on to its website, I found an advertisement for the CW:

> Participants of City Walk get to go on a journey through the enchanting streets of the inner city of Paharganj and the area around New Delhi railway station, led by a child who was once living and working on the streets.[12]

Here, there is little or no insistence that the topos of Pahar Ganj is anymore 'real' or 'authentic' than what the CW visitors might experience in a wealthier part of the city, though it apparently is 'enchanting', and it is instead the past of the guide that is supposed to authenticate the experience and make the visitors rediscover this already very discovered area. How this is to take place is also described:

> The confidence and witty smiles of these guides have little trace of the years spent on the streets of New Delhi. The past is however, very much present on the Walk. For the guides, it is a walk down the memory lane, the places held dear to them and how they faced the odds to survive to be where they stand proudly today.[13]

And as the guides in the text gaze upon the worn streets of Pahar Ganj as a 'memory lane', they invite the visitors to replicate this gaze and thereby imbue the topos of Pahar Ganj with a meaning they might not have gleamed from it on their own. In the absence of an actual, authentic slum post 2010, the guides-as-former-street-children are positioned as *prisms of authenticity*, through which the topos of Pahar Ganj is refracted, both as a staged backstage and a touristic borderzone imbued with a culture of informal urbanism that tourists can immerse themselves in for a time.

Reading this, I became interested in how the CW was performed on a day-to-day basis and especially whether the CW-guides as former street children performed as prisms of authenticity like the advert implied, and if so, how? I therefore started the main part of my fieldwork in 2013 by attending a CW every day for a couple of weeks and obtained a CW-script that provided the blueprint for how the guides performed. The result is the map of the CW's route in 2013 compared the 2010 (included in the Appendix) and a set of field notes that in time amounted to an ethnographic account of the interaction that typically took place on it. This forms the empirical basis of the following section of analysis.

During the CWs I attended, I quickly found that the website promised a product of authenticity that the CW-guides didn't, and perhaps couldn't, deliver. Rather than taking their own lives as a point of departure they seemed to stick to a script that mostly contained general information about life on the street and SBT's work and only touched upon their own lives when they narrated their 'personal stories' at the end of the CW. If the visitors indeed saw the streets of Pahar Ganj through the eyes of what they believed was an authentic 'former street child', it would have to be in anticipation of the personal story that was narrated after the CW had in fact taken place.

In my field notes, the first part of the CW is described thus:

A) *Meeting point*

The tourists congregate opposite a big sign on the east side of the road that says, 'New Delhi Railway Reservation Centre'. When the group is complete, the main guide asks the visitors to gather around him and starts delivering the first part of the script, which is essentially a welcome and a presentation of himself and instructions not to give any money to street children if the group should encounter any. It finishes with the words, '. . . *so shall we proceed for that?*'.

Some of the bolder guides then waits for someone in the group to say *'yes'*, instead of tacitly taking the tourists' silence to mean *'yes'*. If they do not, the guide will sometimes just look expectantly at them until someone does. He thereby institutes the format of conversation that pervades the whole tour at this early point, by signifying that what might sound like rhetorical questions in fact call for answers along the walk.

The main guide takes the lead walking down Chelmsford Road and turns left down a dead-end street colloquially called the 'car park lane'. The co-guide (usually a trainee-guide) will remain at the rear of the group making sure that no one lacks behind or gets lost.

Waypoint B) *Walking past a group of small idols and animal feeding grounds*

The car park lane is a gravelled lane 10 meters wide and few cars drive here as it leads to a dead end, where only pedestrians are able to cross Basant Road into central Pahar Ganj. The quiet makes it ideal for delivering information to visitors, which the noise of the other stops might make difficult. On the way to the first stop, the group passes an outdoor barber, who in the warmer months props up a chair facing the wall, where he mounts a mirror and proceeds to cut his customer's hair and shave their beards in the open.

(. . .)

Waypoint C) *1st stop. Car Park Lane*

The car park provides a peaceful, if trash-strewn, environment where the guide is free to interact with the visitors. The main guide starts by presenting himself and giving out instructions as to what is going to happen on the tour and how the visitors should behave. Then he introduces SBT and proceeds to tell about street life and what SBT does to intervene. The main body of information on the CW is delivered here.

(Excerpt from field notes 2013)

To sum up, the CW-group convenes and walks towards an area where the guide can explain what is going to happen. The presence of the barber is a sign of informal urbanism, but nothing you wouldn't be able to find in most places in Delhi.

So, what do the guides tell the tourists? The following is an excerpt from a CW-script that explains 'street life' at waypoint (C). It was quite hard for me to obtain, as I had to go through the CW-coordinator, who had to ask the guides many times to send me a copy of the script before I finally got it. The reason, I learned, was that they were actually supposed to write their own version of the script in their guide-training period – a demand that tapped into the construction of the CW-guides as prisms of authenticity, who told their own story, rather than replicating others'. None of them wrote their own script, however. Instead, a version of the script was circulated from older to younger guides as a kind of digital contraband resembling a copied school assignment, which many of them saw it as. The personal touch was reserved for 'the personal story' told at the end of the CW, but as we shall see in Chapter 6, even here younger guides borrowed from older guides when constructing it.

In the following excerpt, I have corrected the worst mistakes but tried to keep the general tone of the script true to its original.

Street life

Have you seen any street children on the street?
  So, can you guess why the children runway from home?
  What is the reason the children runway form home?
  If we are counting there are many reason to the children runway from but today I will describe some specific reason why they mainly runway from.

Main problem is poverty because their family cannot provide them enough food, facility and an education. Some parents are alcoholic they do not take care their children. Physical abuse, sexually abuse and mental abuse also a common problem among those families. They have stepfather stepmother and parents don't love their child and child feels lonely over there and they left their home.

And some time children want to see city life because when they show movie in village side they thought if they join city they can survive better life. When they are come to city they have not any back ground supporter so

50 *Prisms of authenticity?*

they have not any option that's why they survive their life on the street. On the street they have not more opportunity how survive their life on the street. So mostly children doing rag-picking, shoes shining, cleaning car, begging and working in small restaurant.
(...)
As I told boys doing many thing job on the street they usually earn about 200 hundreds rupees per day and whatever money they earn they spend every day because they have not any bank account for save the money for the future. If they put it in their pocket and at night time when they go to sleep anybody can come and steal from them.
So can you guess where they spend this money?
Not food. Food they get from religious places like Gurudwaras.
Mainly they spend money in two ways Entertainment and drugs.
About the entertainment, every Friday in India released new movie in cinema so at morning time they wear new clothes, and nobody recognize they are street children, they have to go and watch the movie because there environment is darkness. Over there they can sleep easily and they can take drugs easily because on the street live illegal so when policeman watch them so beaten by the police. So their friends suggest them if you take drugs after you will not feel pain body.
Do you know which type of drugs they are take?
If you know glue, white out, whitener and many others. Mostly street children are taking glue because it is very easy to buy any general store and it is very common drugs what they are taking on the street, if they are living with friends they also spend money on expensive drugs like cocaine, hashish, alcohol and who are older they spend money in prostitution area and mostly street children they spend money on gambling like playing cards.
Do you have any question about the street life?
(excerpt from CW-script 2013)

What position is the CW-guide speaking from here? What practices is the script validating and what authority do the CW-guide assume in order to give it a sense of credibility? The guide paints a general picture of what it means to live on the street, but while the CW-visitors might infer that the guide has in fact experienced it first-hand, based on promotional material that represent him as a prism of authenticity, the guide doesn't actually make that claim.[14] He talks about street children in the third person, which implies that he distances himself from this category within the co-performed space of the CW, but he simultaneously implies that he is able to speak for them. Using the conceptual framework of Chatterjee from Chapter 1, I think of the CW-guides in this respect as 'liminal subalterns' – perpetually on the verge of complete resocialisation but still not so completely removed from the category of 'street children' that they can't make a living representing them.

Saving money and/or owning possessions is described as downright dangerous on the street, as it'll get you mugged while you sleep, and so street life in a sense teaches street children not to plan ahead. This ingrained lack of foresight is something the NGO must then attempt to 'unteach' the children, because without a sense that you can make tomorrow better than today by planning ahead, there is little incentive to undertake an education and enter into the obligations that come with being part of a society. The fact that the guides are making a living speaking (more or less) coherent English to tourists is proof that they have moved on from this irresponsible behaviour years ago. The discourse of what pleasures and perils street life contains situates street children as objects of pedagogy, who must be retrained by the NGO to plan for the future instead of living hand to mouth. To the extent that the guide is talking about his former self when talking about street children, the guide also becomes an object of this pedagogy, perpetually on the verge of being resocialised but not quite, since this would presumably make him unable to represent himself as one of the street children. On the CW, he is thereby both situated as the object and the agent of this pedagogy as he shows how he has improved himself and vows to improve himself further.

This discourse of pedagogy can also be seen in the guide's instruction that tourists shouldn't give any money to street children, as they allegedly won't use it to buy essentials for their survival, but, rather, objects of frivolous fun (cinema, gambling) or licentious objects or services associated with transgressive behaviour (drugs and prostitution). This licentious behaviour, of course, resembles the behaviour of a section of the tourists going to Pahar Ganj, according to Hudson (2010, 374) and Davidson (2005, 48) quoted earlier, and the double discourse of the desirous depravity or lamentable deprivation identified as a feature of the 'slum' in Chapter 1 is thus perpetuated here in a discourse of pedagogical charity. A discourse that is partly used it to market the CW online, partly to validate its purpose afterwards. This is important because as Chapters 4 and 5 show, the legitimacy of SBT as an NGO genuinely helping the children in their care directly reflects the CW-visitors' sense of legitimacy as slum tourists gazing at – and interacting with – former street children.

The guides' position turns the interactions with the CW-visitors into balancing acts. The visitors are invited to excuse the lack of English proficiency evident in the excerpt of the script quoted earlier because the guides are situated as both objects and agents of SBT's pedagogy. The CW celebrates the fact that they have reached their level of proficiency, rather than focus on what they lack. On the other hand, the guides are taught to interact with the CW-visitors in order to make the CW engaging, and as is evident from the field notes, this goes so far as to demand that the visitors answer seemingly rhetorical questions with a 'yes' or 'no' as a kind of mandatory co-performative tactic used by the guide. As we shall see in the following, this insistence on a conversation controlled by the guides evolves into fully fledged 'guessing games' as the CW progresses, which sometimes doesn't quite work to the guides' advantage.

By the promotional material, they are situated within the co-performed space of the CW as liminal subalterns that refract and project their authenticity onto the topos of Pahar Ganj, and the CW-guides thereby inhabit a position somewhere between sight and guide. They guide visitors around Pahar Ganj, but they are

52 *Prisms of authenticity?*

Figure 2.2 The 'God Lane', point (g) of the new CW Route

in a sense its biggest attraction. If that is the expectation CW-visitors also have, then they might be disappointed that the guides aren't able to present their own version of the script, where the abstract knowledge of how street children survive in Delhi is exemplified through these prisms. But if they accept the premise that the CW-guides are instead dispassionate dispensers of abstract knowledge about street life, containing facts, figures and statistics, and are furthermore coaxed into answering questions, then they might answer within discourses that the guides haven't yet learned to operate within.

The guide's question, '*What is the reason the children runway form home?*' (*sic*) from the earlier excerpt, isn't meant rhetorically but, rather, as an interactive interruption in the stream of information. But if it is taken as an invitation to make complex statements about for example the interrelation between governmental mismanagement, the structural violence of drought and individual acts of violence perpetrated against innocent children, then the guides either simply don't understand it or won't be able to continue the conversation in a way where they retain the authority as guides.

There are times, however, when the guessing game will enable the guide to perform an identity as a local with an embodied, subaltern knowledge of the area that exceeds the visitors'. One example is in what is colloquially known as the 'God Lane':

G) 3rd stop. 'God Lane'

A lane 1 meter wide leads from the relatively open space of point F) towards the main bazaar at point F). It is a 50 meter walk where the visitors have to turn sideways in order to let oncoming pedestrians pass and some visitors visibly feel it to be somewhat claustrophobic and disorienting, as very little light finds its way down to the bottom.
(. . .)
The group lines up at the entrance of the lane along one wall and proceed to look at the 20×20 cm. mass-produced tiles fixed to the opposite wall depicting Hindu gods, Jesus and the Kaaba in Mecca, thus representing the three major religions in the area.

(Field notes 2013)

And here we switch to the script as the guide asks,

Can you guess why these are here? Do you have any idea about it?
(Visitors guessing)
Okay don't be confuse I am going to tell you. Some years ago people are using this wall as toilet. So now they don't do toilet because god watch them and give them punishment. Even if you right here if you do toilet here 500 rupees fine but it is not work and if put all religion gods pictures it is work as you know India is fully religious country. Who are the atheist they don't believe any gods, but they respect all gods, that's why.

(CW-guide script)

So, the tiles act as urination-prevention measure, and even if a prospective urinator doesn't believe in the gods pasted on the wall, the imagined wrath of religious inhabitants or passers-by is apparently enough to quell the urge to urinate here. In my field notes I wrote the following about the exchange:

The God-Lane-quiz is usually a hit, as CW-visitors react positively to the answer. The quiz offers a potentially embarrassing moment for the CW-visitors because the only answer most of them can come up with about why they are there is, '*maybe the inhabitants think the gods on the tiles protect them*'. This tends to be offered tentatively by CW-visitors, who seem to feel uncomfortable to imply that the local conception of God(s) is so directly interventionist and thereby simplistic to their mind. They are happy to hear the guides' explanation, that the pictures of gods are put up as an anti-urination-measure, because it gives the sociological explanation that removes the charge of superstition from the locals.

The very few Indian visitors on the CW of course know the answer but usually recognise the guessing game as a pedagogical tool put in place to help the guides interact with CW-visitors, and they therefore hold their tongue out of politeness, so that the interaction might proceed.

(Field notes 2013)

Here, the guessing game works because the guide's question satisfies the visitors and furthermore supports the distribution of roles they might expect from the advert. The explanation given in the 'correct' answer to the guessing game draws epistemologically on a sociological explanatory model of how religion might be used tactically[15] in urban spaces and this seemingly subaltern knowledge-form trumps non-subaltern knowledge-forms, which might invoke explanations of superstition, but without rendering non-subaltern knowledge-forms meaningless because the correct answer is still sociological in nature. This reinforces the distribution of roles within the co-performed space of the CW and provides the guides with the opportunity to show themselves as successful objects/agents of SBTs pedagogy, and this makes polite Indians refrain from mentioning that this knowledge is actually not considered subaltern in urban India but is widely disseminated and might be put into an even larger context of religious appropriation of public space in urban India.[16]

A similar guessing game is played at the end of the CW, where the guide will ask how old the visitors think he is and then proceed to answer 'no' to every conceivable answer, until he reveals that he himself doesn't actually know, since he was too young to understand the concept before he lost touch with his parents, or perhaps his parents didn't have any means of keeping track of time that way or didn't find it important. The trickery used to 'rig' the guessing game so that there is no right answer is forgiven by the CW-visitors. In an emotional sense, it would be impertinent to insist that there should be a right answer if the guide does not himself know it and is presumably sad about this fact, and in an epistemological sense, the non-answer provides insight into the effects of street children's lack of contextual knowledge, which is a key theme of the CW.

When the guides play the role of dispassionate dispensers of abstract knowledge, some tourists expect complex answers, but the fact that the guides are not able to provide that becomes the very point of the CW in the preceding examples. Weakness is turned into strength. The first guessing game exemplifies subaltern knowledge to the tourists in a way they accept, because it doesn't undermine their non-subaltern knowledge, while the second game shows the evils of ignorance and thereby validates the work of SBT and thus also the presence of the visitors.

Conclusion

Chapter 1 focused theoretically on the show/shield dilemma regarding the representation of the deprived/depraved, illegal(ised), (pre-)modern slum and used Chatterjee to frame the space of agency allowed the contemporary, Indian, urban subaltern as determined by the discourses ascribed validity within the nation's homogeneous time-space, a time-space Chatterjee calls the *Time of Capital*. Chapter 2 showed how the depraved/deprived slum is also discursively produced as a desirous, creative, urban space of informalism, not only by scholars and activists who revolt against the forced homogeneity of the nation's time-space but also by authors of fiction who simply use the slum as an exciting backdrop to their stories, as well as neo-liberal business gurus who see the withdrawal of the state in the slum as an exciting possibility to test ideas of a labour market unencumbered by the regulations normally imposed on it by the state.

Chapter 2 proceeded to an analysis of how SBT refashioned the CW after the Akanksha demolition in 2010 so that it went through Pahar Ganj – an area theorised as a touristic borderzone shaped by a century-long influx of travellers to the area. Some of these travellers ascribe a certain type of authenticity to it, because it provides a frame for them to transgress the norms of the society they are embedded in at home. The CW is theorised as a co-performed space, where the CW-guides are positioned by promotional material as both guide and sight, and they thereby act as prisms that refract and project authenticity onto the space of Pahar Ganj. Observations of CWs reveal that the guides are positioned as liminal subalterns perpetually on the verge of being resocialised but never quite getting there as that would undermine their position as representatives of the current street children and their ability to discursively represent them – and their former selves – as objects/agents of SBT's pedagogy. This is a pedagogy that not only validates the CW-guides' own practices in the show/shield dilemma but also SBT as an NGO working for the betterment of street children, as well as the CW-visitors who rely on the NGO's validity to validate their own practice as slum tourists.

The chapter finishes by showing how the guides navigate their liminal position with limited linguistic abilities by instigating informal guessing games. In the best of cases these games show them as experts on the informal urbanism of Pahar Ganj, as they might reveal hidden details the CW-visitors wouldn't have guessed on their own. At other times they simply expose their linguistic and theoretical limitations. This indicates that the CW needs to be analysed as an affective experience, rather than a purely intellectual one, and the next chapter attempts to forge an analytical framework within which this is possible.

Notes

1 The argument is very similar to Engels's argument represented in the beginning of Chapter 1 and can thus said to lead back a broadly Marxian critique of the division of labour and how it affects the distribution of space, which in an American context has been explored by for example David Harvey (2012), who also draws on Engels.

2 Sundaram directly references the *Politics of the Governed* as part of his framework (Sundaram 2009, 19–21) but more than that, the analytical move of showing how colonial policies were perpetuated by the nationalist governments of independent India is a signature trait of Subaltern Studies seen in both Chatterjee (1986) and Pandey (1989, 2006).

3 Mulk Raj Anand is the author of *Coolie* (1947). The novel is one of the first books written by an Indian in English that became popular among readers overseas and is a highly didactic exploration of the caste- and class-based oppression in India (see also Mukherjee 2000).

4 Especially Nehru's commitment to a modern nation state is shown by Chatterjee, who quotes one of Nehru's early writings: 'There is something very wonderful about the high achievement of science of modern technology'. Furthermore, Nehru rejects the idea that India is 'religious, philosophical, speculative, metaphysical, unconcerned with the world' and with more than a hint to the colonial administration, he goes on to suggests that 'perhaps those who tell us so would like India to remain plunged in thought and entangled in speculation, so that they might possess this world and the fullness thereof, unhindered by these thinkers, and take their joy of it' (Nehru in Chatterjee 1986, 131, 134).

5 De Certeau (1984, 92) argues that the view from above the city, which sees it as a grid, is not an Archimedean point of view, where knowledge is 'discovered' rather than produced, and the city is thus not transparent, though it might appear to be from afar. The controlling mechanism of this perceived transparency is theorised by drawing on Michel Foucault's concept of the panopticon (Foucault 1977, 1972), which is the imagined gaze of the Other imbibed into the modern subject via a structuring of spaces such as prisons, schools, barracks, factories, offices, hospitals and asylums. But whereas Foucault focuses on the panopticon's impact on the individual, de Certeau points out that its existence also entails the fiction that someone somewhere can see 'the whole' and while he believes that Foucault succeeds in finding a language (discourse) to make obvious the structures of power pertaining to the fiction of this totalising view, he doesn't think Foucault leaves a space for that which escapes these structures (Certeau 1984, 47). De Certeau's intervention is thus similar to Said's in that they both look for an opening out of Foucault's very structural post-structuralism, but whereas Said uses Gramsci's concept of agency, de Certeau conceptualises it in terms of 'strategies' as opposed to 'tactics', where 'strategies are able to produce, tabulate, and impose . . . whereas tactics can only use, manipulate, and divert' (1984, 30).

6 Verma's book is for instance cited in Davis's seminal *Planet of Slums* (2006).

7 Apart from the CW, I have been able to locate a number of tours, of which these are a few: Delhi By Cycle, which tours a number of neighbourhoods on bicycles www.delhibycycle.com, and Hope Project's walk in the area of Nizzamuddin, www.hopeprojectindia.org/html/st_tour.htm.

8 Meged's analytical framework draws on de Certeau's division between the *strategies* and *tactics* employed in everyday life (see the earlier discussion), and she uses it to conceptually frame the guided tour as a co-produced performance where both guides and tourists might depart from the strategy of their script and employ tactics to alter the performance, escape from it altogether or undermine the implicit hierarchy inscribed in its dramaturgy.

9 Erving Goffman's concept of 'the backstage' is described in *The Presentation of Self in Everyday Life* as 'a place, relative to a given performance, where the impression fostered by the performance is knowingly contradicted as a matter of course' (Goffman 1959, 69). One example is the division in a hotel between the 'front' spaces open to guests and the 'back' spaces, where food is prepared, stock is kept, laundry is washed, and thereby crucially where the staff can refrain from performing the role they are paid to perform in the 'front'.

Prisms of authenticity? 57

10 The discussion of how to conceptualise the spaces of authentic tourist attractions versus the one about how the space of the nation is imagined runs parallel to each other. Both types of spaces are understood ontologically as envisioned and constructed, but the texts written in the 1970s – MacCannell and Anderson – concern themselves with how subjects interact with these envisioned spaces as categories that are almost nature-given, or at least naturalised in a Foucauldian sense. The texts written in the 2000s – Bruner and Chatterjee – write to an audience familiar with the social constructionist turn of the 1980s, and these therefore examine how subjects shift between different visions of these spaces, either out of necessity or as a form of play.
11 For an elaboration of what this means in spatial terms, see chapter 2 of Sundaram (2009), 'A City of Order'.
12 www.salaambaalaktrust.com/city-walk.asp seen 2/10/2013.
13 www.salaambaalaktrust.com/city-walk.asp seen 2/10/2013.
14 I later found out that while most of the guides had lived on the street, some hadn't. See Chapter 6.
15 In de Certeau's sense.
16 The many cases of Public Interest Litigations leading to bulldozers demolishing *jhuggis* referred to in Chapter 1 has a religious component. In a country with a history of communal riots (Pandey 1989, 2006) one the most efficacious ways of delaying the bulldozers is by sanctifying certain parts of a jhuggi by erecting a mandir (Hindu temple), gurudwara (Sikh temple) or mosque within its premises. Around the time where I conducted my fieldwork, there were groups holding up traffic by performing *namaaz* (Muslim prayer) on the Jangpura flyover to protest the proposed demolition of a Muslim shrine erected without permission in a nearby upper-middle-class residential neighborhood populated primarily by Hindus.

3 Playing with privilege?
The ethics of aestheticising the slum

Whiteness and slum tourism

Tourists in slums are not necessarily slum tourists. In contemporary globalised slum tourism (Steinbrink 2012), tourists from the global North visit informal spaces on the margins of large metropolises in the global South, but the distinguishing mark of slum tourism is not so much that informal urbanism is encountered on a given tour but, rather, that informal urbanism and the inequality present there is the theme. Tourists from the formerly colonising world visiting the formerly coloni*sed* world always involve encounters with informal urbanism characterised by inequality, but in most forms of tourism, and especially leisure tourism, these gaps in capital and power are explicitly naturalised, while they are highlighted in slum tourism. This is not because inequality is only, or even primarily, encountered within informal urban spaces. In an Indian context, it would for example be wrong to assume that the income gap between visiting tourists and workers employed in the informal sector is necessarily greater than the gap between tourists and waiters at a five-star hotel. It is perhaps rather that the slum, as an envisioned topos whose characteristics are under constant negotiation, lends itself more easily to an illustration of inequality. Consequently, the types of tourism studies that utilise a postcolonial perspective to make naturalised structures of inequality visible are thus quite different from the types of tourism studies that use a postcolonial perspective to critically analyse types of tourism that explicitly try to represent the effects of inequality. The following section starts with how the former might be conceptualised and move on to the latter.

Edward Said uses Foucault's theory of the 'gaze' to study how the colonial gaze produces ethnic and racially othered subjects and spaces (e.g. Said 1978), and Urry and Larsen use this to study how the 'tourist gaze' produces some spaces as being outside 'home' and some subjects encountered on holidays as 'foreign' and thereby situated outside the nation, while sometimes tied ambivalently to it by a colonial past and a post-imperial present. Urry and Larsen theorise this as a tourist gaze that reproduces the colonial gaze on formerly colonised spaces and the subjects living there by drawing on colonial fantasies of ethnic, cultural and racial difference between the tourists from the global North and locals (2011, 116).

Figure 3.1 Pahar Ganj bylane, point (d) of the new CW Route

Within the tradition of tourism ethnography, Edensor and Bruner also attempt to inscribe the presence of a colonial past into contemporary touristic practices, though few of the participants acting them out seem to be aware that this past informs the power relations they enter into as tourists. Edensor analyses how colonial narratives surrounding Taj Mahal is inscribed into tourist narratives (Edensor

Playing with privilege? 61

1998, 59–89), but he has little use for an explicitly postcolonial perspective other than to say that it shows that the authoritative voice of the Western male has been undermined and has left space for other narratives (1998, 6). Bruner's postcolonial perspective draws on Bhabha's 'third space' as a metaphor for the encounter between the global North and South, because it allows for a gradual rearticulation of power relations based on ideas of cultural difference in the tourist encounter, rather than the overarching discourses identified by Said (Bhabha 1994, 28–56; quoted in Bruner 2004, 18). He allegedly prefers this focus to Mary Louise Pratt's (1992) emphasis on the unequal power relation inscribed into the (post)colonial encounter because he views them as deterministic, but Bruner thereby perhaps also chooses to disregard the parts of Bhabha's writings that are more pessimistic about the transformative powers of a performative approach, which this chapter will return to in the next section.

Other studies conducted in a specifically Indian context employ a somewhat more developed postcolonial perspective. Hannam and Diekmann (2011, 87) quote S. Mohanty's (2003) assertion that travel writing was an 'apparatus of empire' as well as Mary Louise Pratt's (1992) classic analysis of the power relation instituted by early colonial travel accounts and how they translate into recent forms of touristic representations of 'native women'. Similarly, they show how 'heritage tourism' often reproduces colonial power relations, both when colonial bungalows are restored and rented out to the descendants of the colonisers who had them built and when displays of 'native culture' are fashioned into truncated versions palatable to tourists (Arundhati Roy 1997; quoted in Hannam and Diekmann 2011, 43).

These studies all try to employ a postcolonial perspective that connects the blatant inequality present in the touristic borderzones they analyse to a colonial past, though this perspective seldom occurs to the participants in the tourism performances. But none of these texts engages with a theoretically informed subaltern perspective on what meaning it is possible for postcolonial subjects to produce in the tourism encounter, quite possibly because the type of tourism analysed here doesn't have as its explicit aim to listen to subalterns in the first place.

Types of tourism that try to engage with subalterns and to represent and alleviate their suffering include slum tourism and voluntourism, and they carry with them a different set of problems that are also connected to historical forms of domination and privilege. Hutnyk's important study of the *Rumour of Calcutta* (Hutnyk 1996) for example analyses how the image of Calcutta (now Kolkata) as a city of decay are linked to events like the 1942 famine and the influx of Bangladeshi migrants and how these are refashioned in the myths surrounding Mother Theresa's charity in Kali Ghat as a space where the solitary western saint intervenes in these seemingly locally produced problems.[1] Volunteers from the global North at this charity use these representations to make sense of the mixture of fascination and horror they feel at the squalor that surrounds them, but even though their own position is at times described as somewhat futile in its battle against larger structures of inequality, they seldom see themselves as complicit in perpetuating them with their actions, and if they do, they don't seem to know what to do with the guilt connected to this.

Preliminarily, we might conclude the following: the possible moral problem of leisure tourism to the formerly colonised world is that it naturalises the privilege of tourists from the formerly colonising world and glosses over the historical oppression that helped produce it. The possible moral problem of slum tourism and voluntourism, on the other hand, is that their focus on inequality still gives the former colonisers the relative upper hand and perhaps even reproduce the structures of inequality they sometimes attempt to dismantle.

Judging from current literature on slum tourism (e.g. Frenzel et al. 2012; Frenzel and Koens 2014), it seems that local middle and upper classes usually see slum tourism as more voyeuristic and exploitative than the inhabitants of slums themselves. Some even suggest that Indian elites, with their specific relation to its former colonisers, hold a larger grudge than other formerly colonised populations. Frenzel (2016, chap. 6) compares for example Indian and Brazilian civic elites' responses to tourists' positive valorisation of slum areas in Mumbai and Rio, and he concludes that Indians have been far more recalcitrant in replicating this valorisation. Similarly, while the policing of Rio's favelas has probably resulted in more violent deaths than in India's slums, Rio's emphasis on their favelas as a must-see during for example the FIFA World Cup in 2014[2] (Steinbrink 2014) contrasts with Delhi's demolition rampage up to the 2010 Commonwealth Games, which was largely aimed at erasing the slums from the parts of the city that was visible to the world press (see Chapter 1). To Indian civic elites, it seems, slums are nothing to be proud of, even if money is to be made from marketing them as tourist attractions.

It is unclear to what extent the literature representing the slum as a creative, subaltern space cited in Chapter 2 feels a similar unease that their fascination is shared by tourists, because most of these texts simply don't acknowledge the existence of slum tourism to the areas they describe. One exception is Mitu Sengupta's critique of the movie *Slumdog Millionaire* (Swarup 2009) and its 'Hollow Idioms of Social Justice' (Sengupta 2010, 1) when it suggests that upward social mobility for slum residents goes through an American-style game show such as *Kaun Banega Crorepati* (*Who Wants to Be a Millionaire*)[3] rather than the work of the thousands of non-governmental organisations (NGOs) working in slum areas, for example. She thereby critically engages the armchair tourism to the Indian slum that can be undertaken by going to the movies but not the actual face-to-face encounters.

The history of this resentment towards what might be termed a 'white' fascination with Indian slums might fill a book of its own, but the remainder of this section attempts to provide a short contextualisation and theorisation of how it has functioned historically. In Homi Bhabha's (1998), exploration of 'whiteness' he writes of Foucault that one of his most important insights is that 'the place of power is always somehow invisible, a tyranny of the transparent', and Bhabha uses this to illustrate how whiteness functions as a form of privilege that is powerful precisely because it embodies a naturalised, and therefore invisible, norm.[4] Drawing on Toni Morrison's pivotal essay 'Playing in the Dark' (1992), Bhabha connects Foucault's point to the colonial history of terror perpetrated in the attempt to construct and subjugate what is defined as 'not white', and he thereby points to 'the violence [whiteness] inflicts in the process of becoming a transparent and transcendent force of authority' (Bhabha 1998). He writes within the context of the black slave trade and the plantation labour that haunts Morrison's American fictions, but he provides a sort of translation to an Indian context in the earlier essay 'Of Mimicry and Man: The Ambivalence of Colonial Discourse'. Here, Bhabha's basic idea is that whiteness, in a metaphorical sense of the word, is something non-whites are encouraged to strive for, though they might at any given moment be robbed of this whiteness and be called out as non-whites. Bhabha invokes the colonial era as historical context, where Macaulay[5] argued in his *Minute Upon Indian Education* that English-medium-educated Indian clerks should act as, 'interpreters between us and the millions whom we govern; a class of persons, Indian in blood and colour, but English in taste, in opinions, in morals, and in intellect' (Macaulay 1935). The educational goal they were encouraged to achieve was modelled on the English gentleman, but it was understood that they could never become English gentlemen themselves but were destined to only mimic them in their job as Anglicised 'interpreters' (Bhabha 1984, 128).

Whiteness thus functioned, and functions still, as an ideal that erases all other ideals and thereby becomes *the* ideal rather than *an* ideal. Because of its implicit racial premise, non-whites cannot hope to inhabit it, though it is implicitly understood that they must strive towards it. This dynamic is tied to an exoticisation of India's poverty in ways that sometimes seem strange. The Subaltern Studies Group shows that the arguments against an independent India in the pre-independence era were often articulated within discourses that produced communal discord and oppression as something that could only be held at bay by a colonial government, rather than stemming from a set of social structures that were sometimes encouraged by the British administration because it represented the 'natives' as pre-modern groups fighting each other rather than the colonial administration. Katherine Mayo's highly controversial book *Mother India* (1933) is an example of a text that utilises a similar logic within a humanitarian discourse to argue for colonial rule. The book sets up female empowerment as an ideal that functions as Bhabha's conceptualisation of whiteness: *the* ideal always to be strived for, never to be inhabited and in the meantime functioning as 'proof' that India can't protect its citizens or govern itself. Sinha's (2006) chronicle of its reception shows that while it became very influential in England as a view into the underbelly of India, prominent front runners in the growing independence movement like Mahatma Gandhi received it as an example of the colonisers' almost-pornographic preoccupation with the filth of India, paired with the simultaneous lack of acknowledgement that they were complicit in producing it.

It is true that a representational strategy similar to Mayo's was also used by a British elite to justify its governance of British subalterns living within the colonial centre. Seth Koven's book on *Slumming* in Victorian London (Koven 2004) paints a picture where desirous depravity projected onto London's underbelly were to be cured by a stronger commitment to Christian morals instilled in the working class by the humanitarian work of the upper class. Comparing Koven's book to Sinha's book, the crucial difference is that the heathens of Shoreditch, London were perceived as white, whereas Mayo's protagonists were constructed as racially and ethnically other, and their perceived depravity thus seemed to become a function of this ethnic and racial otherness. The blatant racism visible in this and many other texts of the period shows it not only as a crucial component of colonial oppression but also as one of the main reasons why members of the Indian elite felt it necessary to fight for independence at all. Reading for example Mahatma Gandhi's autobiography (Gandhi 1982) as an account of the road to Indian independence, two things seem clear. The Gandhis, Jinnahs and Nehrus[6] who spearheaded Indian independence were all former members of the class of English-educated mimic-men[7] that Bhabha refers to, and in the case of Gandhi, it was his experience of racial discrimination, when he was posted in South Africa as a London-educated colonial officer, which pushed him towards Indian nationalism, because he realised he would always be a second-class citizen within the British Empire, even if he retained his comfortable job as a lawyer.

Today, reactions to slum tours seem to indicate a suspicion among the Indian middle and upper classes to foreigners' preoccupation with how India treats its subalterns, and while the historic deployment of the 'white' subaltern perspective described earlier is part of the reason, it also remains a sore point because there are open questions of whether the living conditions have actually improved for subalterns in India after independence and to what extent racism has been eradicated? Bhabha's conceptualisation of how whiteness functions as a universalist ideal that cannot be attained for certain populations echoes Partha Chatterjee's critique of Benedict Anderson's idea of the homogeneous space-time of the nation as on the one hand the *only* space-time of the nation, and simultaneously a space-time that subalterns cannot hope to inhabit (Chatterjee 2004, 5). This is not incidental, as one of the key thoughts of the Subaltern Studies Group was that the elite of independent India reproduced the colonial structures of government when they gained power in 1947. Chapter 1 has already described how the basic relationship between ruler and ruled was perpetuated in the violent process of urbanising post-independence India, where the naturalised, transparent norm of the ordered, modernist city resulted in frequent clearings of villages that lay in the way of the ever-expanding city. But what happened to conceptions of whiteness? When Indian independence gave power to the 'mimic-men' of the British empire, the 'racial' ideal of whiteness was vehemently opposed by the Congress Party, both in its hierarchical distinction between Indians and Englishmen and in its distinction between castes (Dirks 2001). The role that politicians such as B.R. Ambedkar[8] were allowed to play in the first government after 1947 attests to this fact, but Ravi Sundaram still uses the term 'urban apartheid' (Sundaram 2009, 55) to describe the spatial ordering of Delhi from 1960s onwards. To what extent is it accurate to invoke South Africa's violent regime based on skin colour as a metaphor for the policing of space in Delhi?

The discussion of how casteism relates to racism has been an especially contentious subject since World War II, when the formerly colonising nations in the

Playing with privilege? 65

post-Holocaust era began to see racism as uniformly evil and, in some cases, as a cause for international intervention. Berg (2017) shows how an attempt to align casteism with racism at a world conference in post-apartheid South Africa in 2001 led the Indian government to speak in favour of a stricter delimitation of the concept of 'race' to explicitly exclude caste from it. The Indian delegation thereby sought to prevent the international community's scrutiny of caste discrimination in India, as well as any subsequent interventions against it, and it did so by utilising a discourse that spoke against neo-colonial interventions in national matters. The resistance towards the exoticisation of India's social problems, exemplified by the anger towards Mayo's book, thereby seems to have proliferated into a generalised resistance towards outsiders voicing critiques of how India's social problems are handled. Interestingly, Berg also notes that the interest groups trying to represent the victims of casteism in India at this conference, by aligning themselves with the universalist principles of outsiders, were discursively produced as outsiders themselves by the Indian delegation. They thereby encountered a dilemma similar to the one outlined by Chatterjee in Chapter 1 about the conditions of speaking as a subaltern.

In terms of how racism functions at a day-to-day level in India, recent studies show that it is still prevalent, both in the shape of discrimination between castes and towards populations that were given a lower status in the British Empire, like sub-Saharan Africans who have been targeted in racist attacks in later years (Jain 2016). The caste system does bear resemblance to the stringent, but false, notion of biological race that served as the ideological foundation for apartheid in South Africa, but specific acts of discrimination in India rather seem to be structured along intersecting lines of marginality based on sociocultural and economic parameters such as 'race', caste, class, gender, sexuality and so on. This means that not all marginal positions create marginalised subjects, and Puri's (2017) recent study shows for example that the stigma of homosexuality is mitigated or exacerbated by configurations of 'race', caste and class in the policing of same-sex couples in Delhi so that fair (whiter), upper-class, upper-caste homosexuals[9] might face very little resistance or harassment in India, just like comparatively wealthy, black American tourists might face very little outright racism when visiting India.

Conversely, when population groups are seen as undesirable by the upper classes of urban India, they might be identified by markers of ethnicity and class. Baviskar remarks for example in her study of the resistance towards cycle rickshaws in Delhi that

> [t]he men who pull them are marked by their Bihari and eastern Uttar Pradesh accents, and quickly identified as migrants. Regulating rickshaws in the name of curbing urban congestion is, then, also tinged with anxiety about the influx of migrants into the city, and fears about the collapse of civic infrastructure under their weight.
>
> (Baviskar 2011, 409)

66 *Playing with privilege?*

Migrants from the poorer states to the east of Delhi are singled out as foreigners to the city by virtue of ethnicity, though this ethnicity is primarily seen as a problem when upper- and middle-class citizens have to decide whether these groups should be viewed as legitimate inhabitants to an overpopulated city with lacking infrastructure or as interlopers who might be stripped of their livelihood and expelled from the city.

Later this chapter returns to the former street children working for Salaam Baalak Trust (SBT) on the City Walk (CW) as guides, but for now it is interesting to see just how fundamentally they are distancing themselves from the type of Indian upper- and middle-class sensibility described earlier. The guides are not only themselves poor migrants from the states to the east of Delhi; they also represent a larger body of current street children who have migrated in a similar fashion, and they do so to representatives from the global North by performing a somewhat exoticised version of a slum that the upper and middle classes don't even think should be allowed to exist.

To sum up, the colonial history of domination-through-representation forms the backdrop to the Indian resentment towards forms of tourism that try to engage with subaltern subjects in India and alleviate their suffering, partly because this project of alleviation is tied historically to colonial domination as it was instated as an ideal that India could never live up to by virtue of its non-whiteness and that the colonisers didn't live up to, because they primarily served their own interests. This resentment seems furthermore to be exacerbated by the fact that the elite of independent India in many ways hasn't succeeded in alleviating the suffering of Indian subalterns to a larger extent than the colonial powers that went before them.

Privilege and playful abjection on the CW

So far, SBT's CW has been conceptualised as a co-performed space where meaning and subject positions are continually negotiated between the participants, but while observations and interviews might give information about how this space is perceived by its co-creators, it doesn't guide us towards a phenomenology of how privilege actually functions, when bodies with different markers of for example 'race', class, or gender encounter one another in touristic borderzones, especially since the privilege allotted certain markers of identity is perceived as a transparent norm and thus might be imperceptible to the privileged.

In their study of touristic photographs of slums and informal urbanism in South and South-East Asia, Dovey and King (2012) create a framework for studying the ethical implications of employing an aestheticising gaze on slum and informal urbanism. Like Sundaram (2009), they theorise the encounter with informal urbanism as a Benjaminian *'Shock of the Real'*, which *'cuts through ideology'* understood as a *'form of collective* imagination' (Dovey and King 2012, 13–14), and this is linked both to the encounter with poverty, and thus inequality as such, as well as to the labyrinthine quality of informal urbanism, where tourists might lose their spatial orientation and become 'amazed' by the maze of buildings they are situated amongst. Like Dyson (2012) they also see informal urbanism as often associated with constructions of authenticity, and they see it as a space that produces 'intensity', understood as 'affect' in a Deleuzean sense, by framing affluence and suffering together in dialectic images.

Based on this framework Dovey and King conclude that the pleasure of encountering informal urbanism must be reframed as what Burke (1958 [1756]) calls the 'sublime', which is 'the combination of anxiety and pleasure we experience when we encounter a potentially overwhelming threat under conditions of safety . . . [where] . . . the pleasure depends on being safe in this encounter with the overwhelming' (Dovey and King 2012, 12). They depart from Kant's assertion, however, that an experience of the sublime involves a loss of the moral and rational self, which would mean that a simultaneously aesthetical and ethical interpretation of the tourist's encounter with informal urbanism is impossible, and they move instead towards Lyotard's idea of the avant-garde as a moral encounter with what lies at the edge of comprehension.

Dovey and King's approach is efficacious in that it expands the limited repertoire of affective approaches found within some discussions of slum tourists' voyeurism. Selinger and Outerson (2011) for example conduct one such discussion from a purely philosophical standpoint and conclude that insofar as all tourism is inherently voyeuristic, then slum tourism is no different. But their analytical scope fails to take account of the historical battle over what the sign 'slum' signifies – especially in the context of colonialism discussed earlier – and how this genealogy of signification informs actual meetings between slum tourists and urban subalterns. Dovey and King's approach takes this dynamic into account and enables a reading such as the one undertaken earlier. Yet, their approach doesn't guide us towards an understanding of how 'racial' and social privilege functions in the simultaneously aesthetical and moral encounter with poverty.

On her blog called *Feminist Killjoys* Sara Ahmed elaborates on what she calls *sweaty concepts*, which are concepts we might fashion out of the lived experience of how bodies occupy socially defined spaces. She starts her argument in the abstract space of 'social forms':

> Understanding sexism and racism is about working through how social forms are stabilised; it is working out how possibilities are eliminated before they are taken up. . . . If we start by describing these mechanisms for stabilisation we will be turning things around.
>
> (feministkilljoys 2014)

The invisibility of the implicit norm identified by Foucault and Bhabha in the above is conceptualised here as social forms that are stabilised, and in this stabilisation certain possibilities are marginalised as abnormal and even impossible. Ahmed encourages her readers to chronicle and critique how the invisible norm became the norm and what possibilities it forecloses, and thus her errand seems to continue the tradition of Foucault and his followers. Later in the text, however, she links this basic dynamic of social forms to how privileged bodies move through space by conceptualising social space as 'holding patterns'. She writes,

> Accounting for a holding pattern begins with describing that pattern. And patterns are often the things that do not come into general view. We need 'sweaty concepts' because we need more descriptions of the patterns that are obscured when bodies are received by spaces that have assumed their shape.
> (feministkilljoys 2014)

Social forms produce privilege and the mechanisms that stabilise this privilege produce social spaces, where certain bodies displaying certain markers fit in, while others have to work to do so. This is both a metaphor for how bodies are constructed in the symbolic order – that is in discourse – and a framework for understanding how these discourses physically affect how concrete places are, first, shaped architecturally and, second, policed by a number of different actors to accommodate certain bodies while rejecting others. The bodies that are not rejected simply fit and are thus doubly privileged: by fitting and by not having to even be aware that there is something one should strive to fit into. What Foucault conceptualises as the transparency of power at the discursive level thereby translates into an imperceptibility of privilege at the bodily level. Ahmed tries to combat that imperceptibility by encouraging scholars to develop concepts that carry the marks of how we rub against certain social spaces or observe others doing so in our inquiry. The 'sweat' of these concepts must be a visible genealogy of the privilege encountered in our inquiry, understood as a tacit acceptance or rejection of certain bodies.

Employing Ahmed's perspective, the literature cited in Chapters 1 and 2 critiquing the 'urban apartheid' of Delhi is full of examples of not only how the city has been shaped to accommodate certain bodies rather than others but also how certain practices that some bodies are forced to engage in by circumstances are illegalised or made impossible within social forms that have been concretised in the physical shaping of the city. Baviskar (2011) shows how the widening of roads privilege motorised transportation and turns the mobility of cycle rickshaws and even pedestrians into a problem. Ghertner (2011) shows how malls privilege consumers and thus turn large parts of the city into spaces where only bodies that look like they possess the buying power to participate actively in this economy fit in. Certain bodies inhabiting Delhi are received by it, while others are turned away at the door,[10] and we thereby arrive at a phenomenological version of Chatterjee's framework, where he sees the contemporary, urban subaltern as navigating strategically in relation to the space-time of the nation, which is, in fact, the space-time of capital.

Applying this perspective to the study of globalised slumming reveals a distribution of privilege that is less straightforward. On one hand, tourists from the global North leave their comfortable hotels to step physically into a space they haven't learned to navigate and which hasn't '*assumed their shape*', as Ahmed would have it; on the other hand, they are sure to do so with a guide that might teach them to navigate it, and they do it as a form of play they can quit if they tire of it. In their book on tourism in India, Hannam and Diekmann reflect on this dilemma and conclude that a straightforward condemnation of slumming as a form of 'abject cosmopolitanism' (Nyers 2003) is too neat, as they believe it would imply that slum tourists look down on the slum's inhabitants in their abject poverty. They write that

> this dualism between the tourist who gazes and the object of poverty is too neat. In fact, our research has found that it is rather the tourists who feel abject and out of place in the slum on tour, while the seemingly 'abject' host is in place and not abject at all. People who live in so-called 'abject poverty' seem to get on with their lives; it is the figure of the emancipated Western tourist who is abject in relation to poverty.
>
> (Hannam and Diekmann 2011, 38)

As mentioned in Chapter 1, there seems to be little evidence that slum tourists look down on the guides and local they visit, and Frenzel (2012, 60–1) even goes so far as to problematise the 'same-ing' (reversed othering) that takes place when tourists are too eager to align themselves with inhabitants of slums in an act of solidarity that erases the differences between them. Bringing in Ahmed's perspective, we might say that these tourists don't really work hard enough to make their own privilege perceptible – perhaps because their concepts aren't 'sweaty' enough.

Hannam and Diekmann's suggestion that the tourists are more out of place and abject than the locals seems to point to the same dynamic, but perhaps this entails more than a mere reversal of positions so that it is suddenly the tourist that is abject? How does abjection even function at a bodily level, and how does it relate to the imperceptible privilege of whiteness and the pleasures of encountering the slum?

During the first months of my fieldwork, where I produced my initial ethnographic account of the CW, the questions raised earlier became relevant in relation to how CW-visitors were positioned socially and physically when encountering instances of informal urbanism on the CW. The following excerpt is from the latter part of the CW:

L) *Food stalls*

After crossing the road, the group enters a narrow lane of one-storey houses, where food is prepared in the open. Great grimy pots of boiling oil placed directly on LPG-canisters hold frying samosas and pakoras, and pieces of marinated meat skewered on seekhs are hanging from the walls waiting to be grilled. Garbage is generally thrown into the lane to be picked up by one of the many garbage-collectors in the area, and it is therefore common to see street dogs, cats or even the occasional rat looking for scraps there. Some visitors not used to this squirm as they move through the lane and I've heard some guides introduce the lane this way: *'If you eat this food you will get Delhi Belly.'*, meaning diarrhoea.

According to the guides this place is also generally less welcoming towards the walk than on the other side of the main bazaar, as some locals will sometimes loudly say things like *'What are they doing here?'* or *'Why don't they show them something nice, like the Red Fort?'*.

M) 6th stop. Pottery market

As the group emerges from the lane, they encounter a large trash-room right next to a pottery market where a group of Rajasthani women sell their men's produce. The trash-room is a 10X10m. storage space for the decaying household refuse from the area, which is emptied once a week by large trucks, but it is also the location of a number of recycle shops, and since they are located close to the station, this is the place where tourists are most likely to be accosted by street children. Some of them are selling the trash they have been gathering at the station where they generally also sleep, while others are carrying small bags of Tip-Ex correction fluid that they breathe through, while making more or less coherent pleas or comments at the tourists, who mostly – in spite of the guides' instructions – do not know how to turn them away.

This is poverty at its most smelly. The preceding account was written towards the end of February, but as the year progressed the smell from the trash room became ever-more pervading, and by May the 40+ degrees Celsius of 11.30 a.m. meant that the garbage could be smelled from the food stalls, causing it to mingle with the pungent smell of marinating meat and *chole bhatura* (curry). It could thus well be argued that this is the point on the CW, where abject poverty is experienced in its purest form, as the sticky, smelly reality of garbage-recycling, invades the bodies of the tourists via the nose and eyes and forces them to experience it. But how might we conceptualise this abjection?

In her chapter on 'the abject' and the body's reaction of disgust, Ahmed (2004, 86) uses Tomkins's study of the almost automatic reaction of revulsion the body can have towards certain objects, and she concludes that disgust takes place as a violent pulling away from something that is perceived to be so 'sticky' that close proximity might lead to transfer and thus 'pollution'. The relation between disgust and abjection (theorised by Julia Kristeva) is that the abject – understood as the horrible – threatens to invade the body or perhaps insidiously has already done so, so that the experience of abject disgust is the confirmed suspicion that the body has already been invaded, contaminated or polluted.

It is important to remember in this context that the body is not really separate from the world but is sustained by a circulation into and out of it of for example food, water and air, without which the body dies. Not only do we exist physically by continuously eating the world around us, thus making it a part of ourselves; we also mate and thus create other bodies from our own, by transgressing the surface

Playing with privilege? 71

of each other through sex. Desire (towards sex, food or something else entirely) can be seen as an urge to increase the circulation of certain objects into the body and make them one with it, where disgust is the urge to block this circulation. But even when desirous food is consumed and some of it becomes part of the body, other parts are expelled as faeces that very quickly become disgusting to us. Desire and disgust are thus linked intimately as two opposite reactions to the same circulation in and out of our leaking bodies, and as emotions they illustrate that our bodies are not separate from the world; rather, this separateness is performed when we either pull away from something in disgust or give in to the desire to ingest it.

Walking past the food stalls on the CW, the proximity between potentially desirable food and potentially polluting garbage speaks to the bodily circulation of desirable objects entering the body and disgusting objects leaving it, and the abject thought of a stomach upset suggests itself, as visitors are able to smell both the pungent food and the putrid garbage simultaneously. Here, the linkage between desire and disgust can also be seen in our push–pull relationship with disgusting objects, or as William Ian Miller formulates it, '[w]e find it hard not to sneak a second look or, less voluntarily, we find our eyes doing "double-takes" at the very things that disgust us' (Miller 1997, x in Ahmed 2004, 84).

The signs that signify objects of disgust/desire circulate through our bodies as utterances in conversations we have with others, with texts or privately in our minds. This circulation of signs runs parallel to the circulation of the physical objects the signs refer to, and Ahmed sees this circulation of objects and signs as what constitutes the development of our very subjecthood. We eat, defecate and have sex in a dialectic relationship with the language we gradually develop to describe these actions, and the emotions attached to these acts become linked to the words that describe them – sometimes to such an extent that words describing something can invoke similar reactions as its physical proximity. The earlier description of the CW might have invoked feelings of disgust in some readers, even though only signs referring to disgusting things, rather than the things themselves, were encountered in the reading. But this again means that reactions of desire/disgust, while almost automatic and instant in the reaction of the body, are nevertheless taught to the body by what it encounters over time. Consequently, desire/disgust ascribed to certain objects and can be re-ascribed over time so that the body 'unlearns' reactions of disgust and desire.

At an earlier point on the CW, the guides show visitors a quite different, and less abject, example of how recycling might take place:

D) *Walking past heaps of garbage*

The group passes what effectually serves as a garbage dump, where some of the older trash collectors often work sorting profitable items from unprofitable ones. It then crosses Basant Road and enters one of the larger lanes of Pahar Ganj proper, passing one-room-shops where the visitors might stop to buy cigarettes or water in the summer. Here, some of them will also encounter for the first time the congestion caused by the intermingling of cows, dogs, pedestrians, two-wheelers and carts drawn by humans or bullocks, which some of the visitors find confusing at first.

E) *2nd stop. Recycle shop*

On the corner of two small intersecting lanes lies one of Pahar Ganj's many 'recycle shops' that buys material collected by what the guides call 'rag pickers'. This shop consists of a single room containing a large metal scale hanging from the ceiling as well as piles of recyclable material that spills out into the street covering an area as large as the shop itself. The trash mainly consists of bottles, plastic and paper and apart from stale chapattis (neatly collected in plastic bags and resold as cow fodder) there is no food waste and the place thus smells much cleaner than the garbage heap at point D) or the large trash room next to the pottery market, point M).

It is not a shop that many street children frequent, as they mainly collect trash on the station and this shop is not the closest one to that area. It lies conveniently on the route however, and the guides have an understanding with the owner that if the visitors do not take pictures of him, he does not mind the daily interruption of 1–4 groups looking at him and his shop – in fact he said in an interview I conducted that he feels proud that tourists would include his shop as a sight to be seen on their trip of India.

In this encounter with garbage, its sticky, smelly quality is toned down, partly because no decaying matter is physically present but also because the stop focuses on how useful street children are to the city when they help convert garbage into '*reincarnated commodities*' (Sundaram 2009, 12). If CW visitors were to follow the logic of pollution and stickiness that Ahmed develops, then one reaction towards garbage collectors might be to deduce that if the garbage is sticky, so are the people collecting it. But Hannam and Diekmann's claim that 'the seemingly "abject" host is in place and not abject at all' speaks to the fact that the stated aim of most slum tours is to represent the slum as a space where the guide fits and where it is safe for tourists to let the circulation into and out of the body remain in place, and Hannam and Diekmann's simultaneous claim that 'it is the figure of the emancipated Western tourist who is abject in relation to poverty' expands this point, as it shows the tourist's possible reaction of disgust as problematic.

For tourists to deliberately bring themselves in a situation where they might be disgusted, and simultaneously shamed for being so, might seem strange, but Davidson's analysis in the above provides a possible explanation when she describes certain forms of tourism to India as 'beyond the boundaries' experimentations' brought about by 'submitting the mind and body to poverty, disease and deformity' (Davidson 2005, 48). The transformative goal of the journey among her respondents is to redefine the boundaries between disgust and desire by interacting with a topos that teeters on the brink between the two, so if the body of the 'emancipated Western tourist' is abject because it approaches the slum prepared

Playing with privilege? 73

for reactions of disgust and desire, then this approach is at least partly voluntarily performed by the tourists.

To teeter on the brink between disgust and desire is not exactly comfortable, but might it be characterised as 'pleasurably uncomfortable'? Ahmed writes,

> To be comfortable is to be so at ease with one's environment that it is hard to distinguish where one's body ends and the world begins. One fits, and by fitting, the surfaces of bodies disappear from view. The disappearance of the surface is instructive: in feelings of comfort bodies extend into spaces and spaces extend into bodies. The sinking feeling involves a seamless space, or a space where you can't see the stitches between bodies.
>
> (Ahmed 2004, 144)

Ahmed's point here about (dis)comfort is similar to her call for 'sweaty concepts' as both perspectives focus on the relationship between social and physical space and how the body is received by both. Bodies who conform to a norm inhabit a seamless space that, like a comfortable chair, is imperceptible because 'by fitting, the surfaces of bodies disappear from view'. To be able to effortlessly conform to a norm is thereby to be privileged to the extent that you forget you are privileged,[11] and to be 'out of place' as a slum tourist in an unfamiliar topos is thereby perhaps not to be robbed of that privilege, but rather to imagine its absence by entering into a topos where privilege is made perceptible by its temporary absence? If so, then this playful abjection can be read as a version of Dovey and King's concept of sublimity presented earlier: a small amount of discomfort that makes visible the comforts otherwise taken for granted.

If we conceptualise the abjection experienced by CW-visitors as a titillating expansion of the signs/objects that might be permitted to circulate in and out of their bodies, then the pleasurable discomfort encountered by tourists has very little to do with the discomfort encountered by inhabitants of informal urban areas, such as street children, as their discomfort usually has to do with the long-term effects of precarious living conditions. While the smell of decaying garbage is abhorrent to guides and visitors alike, it doesn't compare to the discomfort of negotiating the daily violence, abuse and exploitation that street life offers, and while it is perfectly possible to catch a disease from handling garbage barehandedly, the detrimental effects of a 10-year drug habit are most often greater.[12]

Accordingly, two parallel 'tracks' can be identified within the narrative and structure of the CW. A *pedagogical* track narrates the actual hardship encountered when living on the street, while a performative track provides tourists with the bodily sensation of not quite fitting, which is conceptualised here as playful abjection. The visit to the recycle shop provides the link between the pedagogic and the performative, as smelly garbage is showed as an odourless commodity and the collecting of it is portrayed as a way of living that has its perils, though this has nothing to do the with abjection tourists might feel towards it. It might thus seem as if the pedagogical track takes the lead – especially since the collection of garbage is portrayed as benefitting the city, which implicitly casts the garbage collectors in the role of deprived workers rather than depraved scavengers.[13] The following excerpt shows, however, that at the exact point where street children are portrayed as nothing more exotic than workers performing as necessary service to the city under precarious conditions within the pedagogical track, the guides restage the performative track by pointing to photo opportunities of informal urbanism in the shape of jumbled wires hanging from poles and by performing narratives of illicit activity.

E) *2nd stop. Recycle shop*

(. . .)

Instead of taking pictures of the shop, the visitors are encouraged by the guides to turn the camera the other way to take pictures of the impressive net of telephone wires stretching out from one particularly heavily laden pole, and in doing so the guides display a kind of insight into the fascination foreign tourists sometimes have towards the signs of informal urbanism that are typically not present in the cityscapes they navigate in at home.

From my conversations with the guides, however, it seems to me that they learn what to point at long before they learn why. It is only retired guides, who have visited European or American cities, that attempt to explain why tourists to India might find this sign of informal urbanism interesting because of the difference it provides to the visitors' homes.

G) *3rd stop. 'God Lane'*

A lane 1 meter wide leads from the relatively open space of point f) towards the main bazaar at point h). It is a 50 meter walk where the visitors have to turn sideways in order to let oncoming pedestrians pass and some visitors visibly feel it to be somewhat claustrophobic and disorienting, as very little light finds its way down to the bottom. On days where the guides feel up to it, they tell how thieves they have known while living on the street jump between the ledges jutting out from underneath the windows, and how they can then find their way to the uppermost windows of the façade, where the inhabitants have perhaps not bothered to put up the crudely welded lattices that are usually found in front of the windows on the lower stories.

An understanding of the visitor's fascination with informal urbanism doesn't come naturally to the guides, who have never really known anything but the cityscape of the Indian megapolis. This fascination must therefore be learned, first, as a set of sights to be pointed to, such as jumbled wires or stories of dangerous mischief, and later perhaps as genuine understanding of how Pahar Ganj differs from the average cityscape of the global North's metropolises. Chapter 2 concluded that the guides are encouraged to project a type of authenticity onto the topos of Pahar Ganj, where they are situated as liminal subalterns midway between free-spirited street children and resocialised citizens. The preceding shows that this authenticity must be paired with a projected fascination with an informal urbanism they really don't see as anything special because they have grown up with it.

Figure 3.2 The trash room near the pottery market on the City Walk

The pedagogical and performative track of the CW

The discussion of how much emphasis the CW should put on the performative track of playful abjection versus the pedagogical track of sombre reflection, and how much effort the CW-guides should put into learning how to perform the former versus the latter, seemed to be a point of negotiation between different groups of SBT-staff. I recorded these negotiations in my field notes and present them in the following, but the context they appeared in seemed to stem from a deep-seated ambivalence that reached back to the very beginning of the organisation. In 2013, SBT had existed for 25 years, and reports issued to its donors claimed it had then helped more than 50,000 children in various ways.[14] It housed about 500 children in five different shelter homes, of which only about 10 children were engaged in the CW at any one time. These 10 were a part of larger group of perhaps 30 'SBT-ambassadors' representing the organisation at performances staged for donors and international visitors, where they would play, sing, dance and show their artwork and photography. This, however, still left about 470 children staying at SBT who were never invited to perform as liminal subalterns to strangers. Consequently, there seemed to be a large proportion of both SBT-children and SBT-staff who thought of the SBT-ambassadors and the CW as the face of a much larger organisation which did much more important work than merely entertain a few tourists or potential donors.

SBT was founded after the internationally acclaimed director Mira Nair was awarded a Palm d'Or at the Cannes Film Festival for her movie *Salaam Bombay!*

76 Playing with privilege?

(Nair 1988). A number of street children starred in leading roles, and Salaam Baalak Trust (SBT) was initially set up to benefit them, so they wouldn't have to stay street children while the movie made millions. As Mira Nair went on to direct such movies as *Monsoon Wedding* (2001), *The Namesake* (2007) and *The Reluctant Fundamentalist* (2013), her mother, Praveen Nair, who was a social worker by training, quietly expanded the organisation to its current size while maintaining the organisation's connection to the glamour of artistic expression. Today, SBT is run by five trustees chosen from the cultural and economic elite of Delhi, and in 2013 it included Praveen Nair; Sanjoy Roy, who also runs the internationally acclaimed Jaipur Literary Festival; the wife of the editor of the prolific magazine *Tehelkar*; the CEO of a real estate agency; and a rich gallery owner.

Salaam Bombay! translates into *Hail Bombay!*, while *Baalak* means 'boy', and the CW-script translates *Salaam Baalak Trust* as '*saluting [the spirit of the] Child Trust*' (CW-script). The name of the trust obviously resembles the name of the successful movie and thus associates the two, but it seems no coincidence that the meaning of the name celebrates children rather than for example 'saving' them, as in the organisation 'Save the Children International'.[15] It signifies agency and representation. As I conducted my fieldwork, the trustees, especially Praveen and Sanjoy, were quite active in running SBT as an organisation for ordinary children who needed a home for a time, and they attended weekly meetings where ordinary challenges of caring for children in a structured manner were discussed with the different branches of the organisation. But they evidently also found pleasure in staging events where former street children were represented, not only as visible, deserving receivers of aid but also as children who had had the guts to run away from home at an early age and whom the world consequently could expect great things from.

SBT thus seemed to function as an interplay ordinary social work and performance, and the coexistence of the pedagogical and performative track on the CW thus reflected the dual focus of the organisation. The SBT-staff, consisting of social workers recruited from the Indian middle class, and the trustees, recruited from a glamorous upper class, were joined by a third group of volunteers from the global North. In 2006, a British volunteer called John dedicated a year to training the first generation of CW guides, after he had visited similar tours in Rio de Janeiro. As he went home, other volunteers took over from him, such as Jessie Hodges, who even got hired as a volunteer- and CW-coordinator in 2009 and is quoted in Chapter 1, as well as a long-term volunteer called Nick, who still worked with SBT in 2013.

As I interviewed him, he told me how his first encounter with the CW came to shape his life:

> '[I]t was supposed to be the antidote to other types of tourism. It was supposed to be . . . ' yes, I have gone and looked at monuments, yes, I've seen this, and this is from the 16th century, and yes, this is very nice, and . . . but . . . can I please have something that's alive?'. And it was very alive. It

was so alive. It was so vibrant, it was so powerful that I went back to my hotel room and I cried. And that was partly guilt 'cause I was staying at a very expensive hotel, but it was partly that I had been hit by something that was at such a . . . that made me realise how banal the other things were that I had been doing in Delhi, like going to Humayun's Tomb and India Gate and staring at pieces of stone and taking photographs of them.

It wasn't just about giving me information. It was about really quite powerfully changing my perspective or giving me a new perspective, I suppose more accurately, in a way. I really felt (and I don't think that is true for a moment, because street life is more complicated than anything that lasts two hours can tell you about), but I came away from the walk feeling that I understood some of what it was like to be a street child. Now as I say, that's completely arrogant because now, five years later I can look back and see that even then back in the hotel I still didn't understand very much, but it was, you know, I began to feel that I had understood that . . . and that was an experiential thing – is that the word? You know it wasn't *'I know this much more'* it was *'I get this somewhere inside me, I feel it a bit'*.

(Interview with Nick 23:00–28:00)

To Nick, the performative track gave him the illusion that the pedagogical track was more profound than it perhaps was, but in his case it encouraged him to return to the organisation and devote a large part of his life to it. In the five years that passed between the experience and the interview, he had started a British organisation called Friends of SBT, which helped raise money for SBT, and he had privately provided financial aid towards buying and running a 'Volunteer Flat', where I stayed while conducting my fieldwork. Apart from me and the changing population of long- or short-term volunteers, the flat housed Nick and a young man I have called 'Hakim' in this book, whom Nick had developed a relationship to that resembled a sort of informal adoption. Commenting on how and why he had immersed himself in SBT, he responded to my question about who he was and where he lived by jokingly giving his address as 'Gilford, Surrey, England for eight months of the year and Pahar Ganj for the remaining four'.

Continuing in that vein he said,

I'm aware of the irony that most of [what] SBT does is about trying to reintegrate children into mainstream society and most them are runaways. I come from mainstream society and I try to run away from aspects of it. Ahem yes. The irony of that doesn't escape me all the time.

(interview with Nick, b 13:26–14:18)

Nick seems to consciously inhabit an ambivalent position, where he uses the fascination he once felt on the CW to give current CW-visitors an experience in which their playful abjection intermingles with the sense of freedom that comes from not conforming to society, while he is actively trying to get the SBT children he interacts with to conform to society to the extent that they will be able to support themselves by the time they reach their 18th year and SBT can no longer help them. He thereby describes the inherent opposition between the performative and pedagogical track on the CW, while simultaneously describing his understanding of the CW-visitors' attraction towards the CW and the life described there.

This difference of perspective among the SBT-staff, trustees and volunteers translated into disagreements about where emphasis should be laid in the education of the CW-guides. Nick wasn't convinced that the CW would continue to generate funding for SBT in quite the same way, when guidebooks and online reviews from seasoned backpackers started reflecting that it wasn't a slum tour anymore, or really a walk where the guides refracted and projected their authenticity onto the topos of Pahar Ganj, but really more of a visit to an orphanage and an interaction with its present and former inhabitants. As I presented him with my account of the CW and the interviews I had conducted with tourists a few months in to my fieldwork, Nick's response was to draft a paper called 'Developing the Walk' that he wanted to present to the trustees, and he invited me to work on ideas to change the CW's route, so that it might include a new *jhuggi* to visit, and to train the guides to include more of their own stories when delivering the script.

In this endeavour, however, he seemed to encounter resistance from the staff that ran SBT, and especially the then volunteer and CW-coordinator, who was a middle-aged, middle-class Indian woman called Poonam. She was responsible for the CW functioning at a day-to-day level, and if Nick acted as an informal father to some of the CW-guides and CW-trainees, Poonam acted as their mother. She was sceptical of the changes Nick proposed, partly, because she would be in charge of implementing them, but there was also the sense that the tourists' preoccupation with the underbelly of Pahar Ganj and the sordid details of the guides' lives catered to a view of India that exoticised a poverty, which she thought of as a mark of shame.

Towards the end of my fieldwork she found time for an interview, where I recorded her point of view on a number of issues we had discussed in the previous four months. One such was the question of whether tourists should be banned from taking pictures inside the shelter home where the CW ended, because while the Indian Child Welfare Committee had decreed that no single child should appear on pictures that might be circulated online, the CW-guides found that the CW-visitors' interaction with the children at the AASRA Shelter Home went smoother if they could use the non-verbal communication of taking pictures of each other. What Poonam heard from the SBT-staff working at the shelter home, however, was that

> '*we've experienced that it is very difficult to monitor the photographs which the guests on the walk take*' and many times, you know, they get into this explorative phase where they go to the kitchen taking pictures of the kitchen, they go to the staff room and without asking they take pictures of the staff. Why are they so stupid!? And then how do we control that? And then

sometimes the washroom, trying to show the worst washroom they have ever been to. So, why? Sometimes it is glamourizing the poverty.

(...)

Tore: Do you think that happens on the walk as well?
Poonam: I'm sure. I'm sure because a lot of time on Facebook, . . . I have seen absurd pictures, which people think are very good photographs . . . like the wires. The wires, which they find it very interesting. I think again it is a matter of perception. Nick would also find it very fascinating that there is this whole network of wires, which is there and then take a picture of that and post it. I don't see anything glamorous about it.

(Interview with Poonam)

Taking pictures of single children, SBT-staff, the kitchen and the toilet and displaying them as symbols of other people's misfortunes are linked by Poonam to taking pictures of informal urbanism in the shape of jumbled wires. Middle- and upper-class India's distaste for tourists' fascination with slums is thereby perceptible in her as a resistance towards the performative track of the CW, as well as Nick's aim of teaching the guides what visitors might find fascinating.

Conclusion

The chapter began by outlining a genealogy of tourism studies that have utilised a postcolonial perspective to analyse forms of privilege. Focusing on types on tourism devoted to representing subalterns, the chapter moved on to a conceptualisation of the pleasures and perils of aestheticising the slum and connected it to how the privilege of whiteness has been conceptualised in an Indian context. Whiteness has functioned partly as a transparent, naturalised norm, partly as an ideal that non-whites should strive towards, though they could never hope to own it for themselves. This perspective was used to illustrate how equating India with the exoticised slum has functioned as an act of domination, and thus why slum tourism in its later, globalised incarnation, might be perceived as working towards that same end by some Indians. This perspective was then linked to Ahmed's conceptualisation of how the body inhabits spaces and how privilege, phenomenologically speaking, is to fit so seamlessly within a social and physical space that it becomes imperceptible. Analysing the part of the CW where smelly garbage is encountered, the chapter concluded that not 'fitting in' in a slum that you tour, socially and physically, is not to be robbed of that privilege but, rather, to playfully rub against the space it creates. It is a kind of playful abjection.

The analysis then showed that the CW moves along two tracks. A performative track invites the visitors to engage in a type of playful abjection that renegotiates configurations of disgust and desire felt towards certain features of street life, while the pedagogical track communicates to the visitors what perils street life contains. As these perils are structural and longitudinal in nature they have little to do with the playful abjection experienced within the performative track. The visitors are thereby invited to feel something but not what it is like being a street child, though this is exactly what the promotional material on the CW claims.

The chapter finished by showing how the ambivalent relationship between the two tracks is also reflected in the different perspectives employed by different groups within SBT. The steady, pedagogical track of resocialisation is upheld by the SBT-staff belonging firmly to the Indian middle class, who are employed to lead the SBT-children towards a secure middle-class existence like their own. The performative track is marked by a tourist gaze focusing on playful abjection, which is taught to the CW-guides by volunteers from the global North, who remember their own fascination with informal urbanism from their initial encounters with it. While these two groups interacting with the guides don't necessarily agree on priorities, they each seem necessary for the CW to function.

Notes

1. The famine of 1942 was not alleviated by the British administration to the extent it could have because they needed food for their troops fighting World War II, while the British, of course, also played a pivotal part in the partition of India in 1947 that caused the first great wave of migrants from East Pakistan to Calcutta (see also Guha 2008, 83–102).
2. It is worth noting that in the case of Rio, civic authorities also used the excuse of a mega sports event to invisiblise and pacify some of the city's slum areas, but they also tried to 'beautify' and stage some them as tourist sights (Steinbrink 2014), which emphatically wasn't the case in Delhi.
3. The game show plays a pivotal role in the hit movie *Slumdog Millionaire* (Swarup 2009).
4. The classification of Bhabha's own writing attests to this, as his books might be classified as Postcolonial- or Black Literature, while the category White Male Literature only exist as 'Literature'.
5. Thomas Macaulay, member of the English parliament.
6. Mahatma Gandhi, the spiritual father of India; Muhammad Ali Jinnah, the first prime minister of Pakistan; and Jawaharlal Nehru, the first prime minister of India.
7. Bhabha originally borrows the term from V. S. Naipaul's novel *The Mimic Men* (1967) whose perspective on subaltern agency was famously condescending, along with most of Naipaul's *oeuvre* as such. Bhabha critiques this stance and in the following chapter called 'Sly Civility' (1994, chap. 5) expands the idea of mimicry to become a tool of resistance rather than mere submission often utilised in postcolonial literature.
8. For Chatterjee's perspective of the importance and limitations of Ambedkar, see Chapter 1.
9. Like the popular and more or less openly gay TV host Karan Johar.
10. While certain spaces in Delhi rely on architecture to signal which bodies are welcome and which aren't, in a way theorised by for example Foucault's ideas of *dispositives* as physical manifestations of discourses (Foucault 1980, 194–228), most elite spaces are simply guarded by a man with a stick, gun or both, and his sense of the clientele his employers would him like to gain access to works via a recognition of bodily makers of class, gender, age and caste (in terms of skin colour), to name but a few socio-economic and cultural parameters.

11 She uses this to make an argument about heteronormativity, which is the privilege of heterosexuals because public spaces are pervaded by a history of heterosexuality expressed in anything from marriage legislation to public pictures of grandparents of different gender staring affectionately at each other. Privilege is thus a comfort you do not feel because you could not imagine doing without it.
12 See also Chapter 2.
13 This is backed up generally by the discourse criticising urban apartheid, mentioned earlier, but especially by studies such as Kaveri Gill's *Of Poverty and Plastic* (2012).
14 According to statistics released on the occasion of its 25th anniversary that year.
15 See www.savethechildren.net.

4 The affective economy of slum tourism

Tourists' responses to the CW

At the beginning of my fieldwork at Salaam Baalak Trust (SBT) in 2013, I had what might be called an unexpected windfall of empirical data pertaining to tourists' reactions to the City Walk (CW). After each CW, visitors are encouraged by the guides to submit a feedback form evaluating their experience, and I collected a year's worth of feedback forms (2,739 from April 2012 to April 2013) and registered their results in a spreadsheet according to gender, age, nationality and their general evaluation of the CW in five predetermined categories pertaining to the guide's information, communication, attitude, interaction and overall experience. I furthermore identified seven types of recurring suggestions in the section on how the CW might be improved.

Gender and age

Two-thirds of CW-visitors are female, and roughly two-thirds are between the ages of 17 and 35, with a peak at the ages 19 to 21 (gap-year travellers). Given the size of the groups (about 10 participants), this means that most groups contain visitors from a wide spectrum of ages, though still more young than old and twice as many females as males.

Nationality

The CW is visited almost exclusively by people from the global north, as North America, Europe and Australia make up 92% of all nationalities. Of all visitors, 7.7% come from Asia, and 4% of these are Indian. Of the remaining 3.7% of travellers from Asia, 3% come from Japan, Singapore and Israel – which are all relatively wealthy countries for the region. The European visitors on the CW are almost exclusively from Western Europe, with the UK supplying 28% of the European visitors. Denmark supplies 26% of all visitors to the CW, though its population is roughly 11 times smaller than the UK's and the number of tourists from Denmark to India is roughly 28 times smaller than the UK's in the same year.[1] The immediate reason for this can be traced from the feedback forms to the fact that Danish travel agencies, such as Jysk/Visit Beyond, Grace Tours and Kilroy, all include the CW either in their round trips or in 'start packages' for individual travellers.

Evaluation

In the first part of the feedback form's evaluation, 72% tick 'Excellent', in all boxes, or all boxes but one; 25% tick a median of 'Very Good'; and only 3% ticks a median of 'Good'. None tick a median of 'Satisfactory' or 'Needs Improvement', though individual boxes in these categories are ticked. It is virtually impossible to tick a negative box in the form, as 'Needs Improvement' does not explicitly state that the guides should already have learned what they do not master but, rather, that they haven't done so yet. Even so, no one ticks even remotely critically.

The most consistent detraction from the assessment 'Excellent' comes in the category of 'guide's communications skills'. This could indicate that the guides do not always communicate as well as the visitors would have liked, but this does not automatically mean that this is only the guides' problem. Around 58% of all visitors probably have English as their mother tongue, as English is an official language in their country of origin, whereas around 42% consequently do not, and what is sometimes perceived as lack of communications skills on the part of the guides might then also be due to the limited linguistic capabilities of this section of visitors. Second, there is no guarantee that even native English speakers understand Indian English, as it is commonly spoken on the Indian subcontinent. The guides therefore say that they sometimes have to speak a more American-/UK-style English to the native speakers, while employing a less complicated version of English than they are actually able to speak, in order to be understood by some non-native speakers.

Independent suggestions for improvement

I was able to identify seven recurring categories of suggestions on the feedback forms given by 23% of the visitors, meaning that 77% had no suggestions or made suggestions that didn't fall within any recurring categories. The most consistent suggestion made in 10.4% of the feedback forms is to include more 'street life', which probably refers to the fact that this aspect of the CW has been severely diminished since the 2010 demolition of the Akanksha Colony and the ban from entering New Delhi Railway Station (see Chapter 1). Since the CW was in 2013 still marketed in guidebooks and on the Internet as a slum tour – or at least one focusing on 'street life' – the absence of it might have been disappointing to some. Other suggestions include a wish to interact with the SBT girls (2.3%), none of whom were CW-guides, the urge for more information and a longer CW (3.3%) and two almost equally big groups who think the interaction with the SBT-children should be either increased or reduced (respectively, 3% and 2.3%).

The distinct lack of critique in the feedback forms intrigued me. Why did everyone, even visitors who seemed to have trouble understanding what was going on due to communication problems, rate the CW so positively? One obvious factor was the setting where the forms were filled out. The guide being evaluated always remained in the room while this was done, and it always happened right after he had told his moving personal story, making it hard for the tourists to be critical. Even after months of attending the CW, I personally experienced the same difficulty rating them as harshly and honestly as they asked me to, because it seemed quite an achievement that they could perform as well as they did, given their past. Similarly, the reason why the tourists ticked the box of 'guide's communications skills' consistently worse than other boxes could be explained by communication problems but also by the guides repeatedly explaining to visitors that the CW was set up to improve the guides' communications skills. The visitors were thus effectively agreeing with the guides when they said their English needed improvement, thereby also validating the guides' position as objects of SBT's pedagogy.

While registering the result of the feedback forms, I started conducting interviews with tourists that I went on the CW with. I'd pick them randomly by simply asking after the CW if anyone was interested and then take them to a nearby café where I'd ask them about who they were, what they had experienced on the CW and what they thought about it. Initially, my results were as diverse as the group going on the CW, though a few common denominators presented themselves. The group was generally not very well prepared to have a conversation about the CW that went beyond what they had understood of the CW-guide's delivery of the script, because even though many of them could spot the guides' explanations of social dynamics as somewhat simplistic, they hadn't acquired enough knowledge about India, street children, Delhi and so on to directly contradict them. In their conversations with me, they would typically fill these knowledge-gaps by either drawing on pre-existing knowledge derived from their life at home, knowledge about travelling generally or simply ask me what I thought, rather than answering my questions.

I thus had a long conversation with a former drug user from the United States about the drug habits of street children and the benefits of hot yoga, which he described as his new addiction. Did the former street children also do yoga? Another interviewee was a social worker working with children in England, who pointed out likenesses between SBT's work and his own profession, and urged me to fill in the gaps in his understanding of it, due to communication problems. A mother told me at length about the conflict in her native Northern Ireland and how she had taken her daughter to see India because of it, as she herself had travelled there when she was younger and wanted to give her daughter the same experience. These encounters served to remind me that travelling is as much about getting away from something and returning to it, as it is about where you go. Like readers reading novels to gain insight into their own lives by comparison, these travellers were less interested in the perspectives of the people they met than in how their own perspective changed when they gained distance from their lives at home.

Other travellers engaged me in what Bruner has called *tourist talk* (Bruner 2004, 15), where travelling itself becomes the topic rather than where you travel to, and as in Bruner's case, there was a curious sense of wanting to be recognised as a well-performing traveller, which here merged with the urge to be perceived as a proficient interviewee.[2] A long-term backpacker compared the CW to other travel destinations in South or South-East Asia and gave examples of how instances of informal urbanism resembled each other in different countries and of how he avoided being scammed by touts, vendors and tour operators. Two young Danish girls had just arrived and had already been scammed by a tour operator, who had made them pay for a trip to his cousin's houseboat in Srinagar, Kashmir, that they didn't want to go on. They were aware that they had not yet attained the skills of seasoned travellers and tried to perform as proficient interviewees/travellers by instead citing information on India from a guidebook I myself had co-authored a few years previously for a Danish publishing house (Holst and Mukherjee 2011).

The insights gained from these interviews were valid but unsurprising. And while the lack of knowledge explained why visitors had trouble critiquing the guides, it didn't fully explain the overwhelming positivity in the feedback forms. What were the rules pertaining to the expression of emotions within the space of the CW? Was there space to express anger, fear, anxiety, distrust on the CW? To get answers to these questions, I started observing visitors on the CW to see who had trouble fitting in or became emotional in one way or another, asked them for an interview and based my questions on the emotional responses I thought I had observed in them. This group had the same knowledge-gaps as the previous group of interviewees, but their inquiries of me were less about satisfying their curiosity and more about confirming or denying a suspicion that something wasn't quite right.

To get answers, rather than providing them, I would encourage them to explore their sense of unease with me, without providing the answers that they wanted. This put them in a vulnerable position for two reasons, partly because they risked engaging in conjectures that might turn out to be wrong, which might make them appear dumb, but also because if I confirmed their suspicions, they might turn out to have supported an organisation that exploited either the guides, the former street children or the locals in a way as yet unknown to them. It was therefore sometimes hard to establish the bond of trust required to make this experiment work between me and them. I quickly found that tourists who shared certain sociocultural parameters with me opened up more readily. The most important parameter in this respect was language, as some of my most successful interviews were conducted with Danes whose mother tongue I share. Danish is not widely spoken in India, and the conversation therefore felt private even if it was conducted in a public sphere. Apart from linguistic advantages to speaking to fellow Danes, I was also able to forge a sense of a shared moral belief with interviewees whose national or regional culture I shared, which meant that interviewees from other Scandinavian countries also opened up to me, even though we spoke English to each other. Our shared culture helped them feel safe in their exploration of their own discomfort, because even if it resulted in the realisation that they had done something wrong, they sensed that I would at least understand why they would feel what they felt and act the way they did.

In the end, I selected 16 interviews to include in the study, though only three will be cited at length to illustrate different affective approaches to slumming. While conducting the interviews, I started to develop the theoretical framework for understanding how affect functions on slum tours, which is the topic of the next section.

Economies of affect and capital in tourism

Within the humanities and social sciences, the study of affect has in the last decades grown to such an extent that some have begun to refer to an 'affective turn' (Denzin 2008; Clough 2007) that rethinks the relationship between signs, objects and bodies as a reconceptualisation of the 'linguistic turn' of the 1960s. One overview of this field's conceptualisations of affect comes from Ruth Leys (2011), who summarises what she calls a Spinozist-Deleuzean-Massumian anti-intentionalist approach to affect by quoting Shouse:

> [I]t is important not to confuse affect with feelings and emotions. . . . Affect is not a personal feeling. Feelings are personal and biographical, emotions are social, . . . and affects are pre-personal. . . . An affect is a nonconscious experience of intensity; it is a moment of unformed and unstructured potential. . . . Affect cannot be fully realised in language . . . because affect is always prior to and/or outside consciousness.
>
> (Shouse in Leys 2011, 442)

According to Leys, the problem of this 'anti-intentionalism' (Leys 2011, 437) is partly that affect, understood as a nonconscious experience of intensity that circulates between bodies and objects, has to be translated into feelings or emotions in order to be represented in discourse. Affect might be sensed by bodies that are part of this circulation (or registered by special machinery attached to these bodies), but the moment the researcher tries to generate meaning – understood as existing in language and thus discourse – about this affect, it turns into feelings or emotions. This is not a problem for the study of affect as such, but it does mean that the humanities and the social sciences, which broadly study how humans make sense of their existence and organise themselves socially, cannot utilise much of the insights generated by studies of affect because it is epistemologically situated outside this sense-making. Leys second critique of this approach is that it instates the Cartesian body/mind divide that it tries to escape, by theorising the workings of the body as essentially unfathomable to the mind, precisely because affect is understood as a nonconscious experience of intensity.

Yet, the usefulness of understanding affect as something that circulates between bodies parallel to the symbolic order is illustrated in Chapter 3 of this book, where Sara Ahmed's theoretical framework is used to conceptualise a phenomenology of privilege as something more than discursive transparency. This allows the analysis to link the bodily sensation of privilege to the body's reactions of desire/disgust and thus arrive at the concept of playful abjection. But what is the relationship between affect and the symbolic order and how might it be used to study tourism?

Like Leys, Ahmed wants to distance her framework from a theory of affect that 'creates a distinction between conscious recognition and 'direct' feeling' (Ahmed 2004, 46), but as opposed to Leys, who argues that affect and emotions cohere, she begins by arguing that the intellect is constituted emotionally because it is shaped by the interplay of sensations and thoughts. Subjects cannot experience the world without sensing it, they cannot sense without feeling emotions towards what they sense and they cannot think logically without basing their thought process on their emotionally charged senses. Subjects therefore experience the world through emotions and act towards it in the same way. This also means that emotions are not formed in the body and move outside it, nor are they formed outside the body and move into it; rather, they shape what is thought of as 'inside' and 'outside'. To a newborn, it is this exchange of emotionally charged sensations that gradually come to constitute the surfaces of the body, and this means that the metaphor of 'the contagion', where emotions circulate between already-constituted subjects is thereby not accurate either, because it presupposes subjects who are constituted emotionally before this exchange of emotions starts (Ahmed 2004, 16).

Affect is thereby created not by objects or bodies in themselves but is rather constituted by its circulation between bodies and objects, and this circulation is not divorced from the realm of discourse but is an integral part of it. To illustrate the exchange of affect and meaning, she uses the idea of 'impressions' and playing intentionally on both the literal and figurative meaning of the word; she contends that the 'impression' one body might leave on another is physical as well as psychological.[3] From the moment we are born, we *press* against one another and leave im*press*ions on each other when we interact, and thus, we become 'congealed histories of past approaches' (Ahmed 2004, 160) that are continually ascribed meaning. These 'approaches' might well be thought of in everyday terms as encounters or simply conversations between subjects, where affectively laden words circulate between the bodies uttering them. It might even take place in the interaction between reader and text as affectively laden signs circulate in and out of the body in the reading.[4]

But *how* we impress on each other in concrete situations is contingent, because no body or object in itself produces certain affects, as these are produced only in the relation between them. A piece of garbage encountered in informal urban areas is thus not necessarily disgusting, and neither are the bodies encountered there capable of inducing specific emotions in tourists; it is all contingent on the relations between objects and bodies that come into each other's proximity and the history of emotionally charged meanings ascribed to them. Ahmed's argument against the separation of conscious recognition and 'direct' feeling is thereby that even if the body in some cases react before any conscious decision has been taken to do so, and the body thereby might be said to act affectively outside the symbolic order, the reason the body reacts in a certain way has to do with the past circulation of im*press*ions it has been a part of, and these do exist in the symbolic order.

How might this perspective be used in an analysis of slum tours? Ahmed's framework suggests that a translation between affect and emotions has to take place in order to generate meaning about how affect is felt. This act of translation is always done *by* someone, and in Chapters 2 and 3 that 'someone' is me, as the analysis in those chapters is based empirically on my own field notes. When observing my potential interviewees on the CW before the interviews, I was conscious that I might be interpreting reactions into their behaviour that they didn't think they had and affects they didn't feel moving between their bodies and the topos that surrounded them. Asking them how they felt on the CW was thus a way of recording how they performed the translation between affect and discourse, though I, of course, was still the instigator of this process and directed it in certain ways.

Recording this translation, I came to conceptualise the CW as a space where affective laden im*press*ions circulated between bodies and objects, and the subsequent interviews as a space where this circulation was explored retrospectively, sometimes by an imaginative re-enactment performed by the interviewees, when they recalled the exchange taking place on the CW. In my initial observations of them on the CW, I would look for signs that they were, in Ahmed's words, rubbing against the space of the CW in a way that wasn't quite comfortable, and in the interviews, I would encourage them to explore this discomfort and interpret it for me. What I found was that while the performative track of the CW offered an experience of playful abjection, where people unaccustomed to the slum might teeter between desire and disgust in an experience of slight discomfort, the pedagogical track of the CW was situated by the guides squarely in a space of enforced comfort, a space where questions about the moral implications of going on a slum tour were so much 'out of bounds' as to be almost inconceivable. Consequently, very few tourists transgressed the boundary delineating this space of comfort, and when they did, only did so hesitantly and retrospectively in interviews.

Much of this unease had to do with the economic exchange taking place on the CW, which was explained within the pedagogical track of the CW as a question of visitors simply helping where help was needed. To some tourists, however, the question remained whether they were for example donating to former street children or buying an interaction with them? Who was buying and selling which commodities in what manner of ways, and how did the business side of the CW correlate with the humanitarian logic communicated on the CW where help was simply given? To make sense of this, we must first turn to theories of how value is generated within tourism.

In his study of slum tourism Frenzel (2016, chap. 4) asks the apparently simple questions, 'How does a place become attractive, how does it become valuable? And what is the role of tourists in this?' (Frenzel 2016, 64). He starts by making the classical distinction between value and price. In modern economics, understood as a discipline with a set of tools, value equals price, which is to say the value of a commodity is equal to what someone is willing to pay for it. Marx opposes this idea by stating that *labour* is the source of value and that the profit made from selling a commodity is taken from the workers who gave their labour to making it. Frenzel tries to mediate between the two approaches by pointing to how tourist attractions are co-created between labourers and tourists and gradually turns into tourist commodities. Monuments or national parks are not necessarily made, in the first instance, to generate money from tourists but are gradually turned into tourist commodities in a process that involves not only building a tourist infrastructure or marketing it but also other tourists authenticating it as something worthwhile visiting. An authentication that is all the more obvious in the case of tourist attractions where it is the presence of other tourists that is the main attraction, such as in restaurants, cafes or bars mentioned in guidebooks or online reviews as 'the place to be' because everybody else is there. It is thus not only the CW which is co-performed but most tourist experiences as well, and this also means that the value generated from these experiences are co-created. This induces Frenzel to ask provocatively whether tourists as co-creators of value should then also be paid.

To replace the fraught concept of 'authenticity' with 'authentication' has the benefit of focusing on a process that confers value onto a tourist attraction by approving it, rather than stating that it is somehow more 'real'. Frenzel, however, uses

> valorization rather than authentication because the labour theory of value offers the advantage of seeing attraction-making as a *process of activity*, or *value practices*, or *labour*.
>
> (Frenzel 2016, 70 my italics)

Conceptualising how value relates to valorisation, he uses Graeber (2001) to make a distinction between 'value', which is created via what is recognised as labour, and 'values', which are talked about in the sphere where labour goes towards creating things that are not turned into commodities, such as 'housework to hobbies, political action, personal projects of any sort'. Using this distinction, he distances himself from Marx's idea of valorisation as the instance when labour is transformed into commodity, creating a profit that usually doesn't benefit the labourer. Something more than commodities are created in that process, Graeber argues, and not only by what could traditionally be thought of as labourers.

But how might we conceptualise the role affect plays in this process of co-performance? Ahmed calls the circulation of affectively laden im*press*ions between bodies an 'affective economy' because she argues that we can use Marx's model of the generation of surplus value as an allegory of how bodies and objects (and the signs used to describe them) accumulate affect in their circulation. Just like the move of 'money to commodity to money [*creates*] surplus value [*and*] this movement converts into capital [*so does the*] movement between signs and objects convert into effect' (Ahmed 2004, 45). It is for example in the circulation into and out of the body that food (and the sign used to signify 'food') accumulates desire and disgust to such an extent that the mere mentioning of 'shit' (the waste product of food) on this page might invoke a reaction of disgust in the reader.

But what about the circulation of capital? In Marx's theory there is also passion, but it is a passion related to the accumulation of wealth, or as Marx formulates it 'this boundless drive for enrichment, this passionate chase after value [*that*] is common to the capitalist and the miser' (Marx in Ahmed 2004, 45). In her delimitation of her use of affective economies as a concept, Ahmed is quick to point out that she is not interested in this particular passion of accumulation ('greed') but, rather, in the accumulation of all passions in signs by way of their circulation. This book is interested in both, which is to say both how affect is generated in economies circulating signs between bodies and how this affective economy generates capital and vice versa.

Seen in this perspective, the participants of slum-tours accumulate affect by circulating signs such as 'slum', 'street children', 'street life' and so on between the bodies of the participants on the tours, but this economy of affect thereby creates not only emotions but also capital because tourists pay to be a part of this affective economy. The twin economies of capital and affect are thus mutually constitutive of each other, as the capital generated by the guided tours in turn sustains a material framework, which makes the economy of affect possible. The CW can therefore be conceptualised as a co-performed space of negotiation, where affect and capital must circulate in relation to each other in quite specific ways, if affect is to leave its impression on the bodies co-performing the tour and capital is to circulate towards SBT as an organisation.

The idea that emotion generates capital recurs in Hochschild's analysis of the emotional labour of workers in the post-industrial labour market (Hochschild 1983). She combines a post-Marxian discourse on labour (Bell 1973) and a Goffmanian discourse on 'display' (Goffman 1959) with a phenomenology of emotions that stresses their social nature, and Ahmed (2004, 9) later draws on these perspectives, even though Hochschild doesn't conceptualise emotions as physical entities that travel between bodies in concrete spaces.[5] Hochschild's observations of the 'feeling rules' (Hochschild 1983, 56–75) of a workplace reveals what is now widely accepted requirements of certain professions: stewardesses are supposed to be cheery, undertakers are supposed to be sombre, bartenders are supposed to be flirty and so on. To Hochschild this signifies that commodified emotions are bought and sold within a range of industries, and she argues that workers thereby become alienated, not because their bodies are instrumentalised as in a classical Marxian analysis but because their emotions are. By employing a Goffmanian backstage–front-stage metaphor, which invokes a difference between surface acting and deep acting, she shows how the upbeat surface acting of emotional labourers affects their deep acting so that they have difficulty distinguishing between the two after a while.

Thirty-three years later, Johnson (2015) writes from the perspective of a British society where feeling rules have become common within the workplace, and where the kind of co-creation Frenzel describes seems to be the order of the day. Johnson argues that jobs in which emotional labour is de-commodified might, in fact, be exploitative if employers still expect the workers to carry out emotional labour, without giving them due compensation in terms of increased wages.

Discursively producing the good atmosphere of for example care homes as a 'second paycheque' (2015, 117), which implicitly benefits the workers, thus glosses over the fact that the good atmosphere is the product of intense emotional labour on the part of the care workers. Demanding that emotions displayed in staged performances should be 'authentic' (Hochschild 1983, 185–98) is thus problematic not only because emotions are commodified but also because the system of compensation breaks down.

Understood within Ahmed's theoretical framework, this might be conceptualised as the strategically deployed linking and de-linking of economies of affect and capital. Economies of affect produce feelings in customers, while economies of capital ideally compensate workers for the labour of producing these feelings, by sustaining the economy of affect. The paradox is that while economies of affect that seem to be separated from economies of capital are experienced as more pleasurable because the flow of affect is seemingly not contingent on the flow of capital, this seeming separation of affect and capital might also conceal that emotional labourers work much harder at ensuring that pleasurable emotions are circulated and amplified than their 'customers'. Or, to take Johnson's example, the elderly patients at the care home feel better for knowing that the care workers genuinely care for them, but this should ideally result in the care workers getting more money, not less.

Taking Frenzel's provocative question of whether tourists should be remunerated for the value they create when for example visiting restaurants or 'discovering' new parts of town, which in time become tourist commodities, we might ask who facilitates the tourists' ability to perform this 'labour'? Do these restaurants or neighbourhoods employ labourers working towards making the discovery of them a new experience for each wave of tourists, and if so, does the focus on the value tourists create by valorising tourist attractions positively run the risk of de-commodifying the labour of these facilitators?

This question has a direct bearing on how the exchange of affect and capital might be conceptualised on the CW. Like Johnson's care workers, the CW-guides can be seen as emotional labourers ensuring that the right kind of affect is circulated within the space of the CW, and while the CW is clearly 'co-performed', this doesn't mean that everyone within the CW works equally hard at maintaining the comfort of this performance.

Seeing this theoretical framework as a methodology for analysing how economies of affect and capital are balanced in relation to each other within the co-performed space of the CW, we might preliminarily conclude that the guides are in charge of directing the circulation of affect in a way that is satisfying to the CW-visitors, though they must get the visitors to participate in this circulation. Using Frenzel's framework, we might also conclude that while the guides/SBT is remunerated for facilitating this circulation, what is produced is not only a commodity, though a price might be put on it, and perhaps *should* be put on it, in order to avoid the kind of exploitation pointed to by Johnson.

Figure 4.1 The researcher interacting with SBT-children at Aasra Shelter Home

The anxiety of encountering shelter-home children

In the parts of the CW analysed so far, the performative and pedagogical track has been separated into an experience of the topos of Pahar Ganj, characterised by playful abjection, and a dispassionate delivery of information about vulnerable or suffering street children helped by SBT. At the end of the CW, however, the two tracks seem to merge, as these vulnerable street children are encountered in the flesh. This happens at a contact point established at the New Delhi Railway Station and at an SBT-shelter home, and the encounter is described in my field notes this way:

J) *5th stop, at GRP contact point*

The General Reserve Police (GRP) contact point is the oldest in the Delhi branch of SBT and is built on top of the police station attached to the railway station. SBT has in 2013 12 such points that are generally open from 10:00–14:00 and as opposed to the full-care shelter homes are places the children can get access to educational and medical services without having to commit fully to the restrictions imposed on them if they join the trust's shelter homes. Because it is at the station around 1200 children passed through the GRP in 2012 alone, the majority of which was sent back home to their parents.

94 *The affective economy of slum tourism*

Going up the steps, the group encounters the present street children studying on the terrace with the volunteers attached to the contact point, while the inside reveals a small room with a television, where the children can watch movies on Fridays, as well as an office on the terrace overlooking the station, where the children are given medical treatment if they need it and the newly arrived children are registered and offered help by the social workers.

As the CW proceeds to SBT's shelter home, another group of children is encountered:

N) *7th stop. The Aasra Shelter Home*

SBT has 5 shelter homes. 2 for girls situated far outside the city centre and 3 for boys, situated in a 3 km. radius from each other in the centre. Aasra is the smallest and oldest and acts as headquarters for SBT as well as midway station for children whose case is yet to be decided. They could go home to their families, to some other organisation or one of the other permanent shelter homes with more space and facilities and generally they are not meant to spend more than 6 months at Aasra, though scarcity of space sometimes necessitates that they spend longer time there before a permanent solution can be found.

The first thing that the visitors encounter is a welded gate on the landing stretching from one staircase to the next. It is a measure imposed by the Indian Child Welfare Committee, who has wanted to make sure that the children can't leave once they have joined the shelter home, though the children sometimes find ways to escape anyway. Inside, the visitors are led to the activity room, where the children are generally convened at this time of day to greet them before their lunch and a 10–15 min session of interaction follows.

Sometimes a group of volunteers are engaging the children in some activity like drawing or reading, which makes it easier for the visitors to join because they thereby have something to do with the children, rather than merely trying to communicate in whatever Hindi they know (usually none) or whatever English the children know (usually very little). At other times it is up to the guides to start some sort of activity, like dancing or singing, and here some visitors are overly conscious that they do not want to overstep the boundaries of the children, sometimes perceiving an exploitation of them that is neither felt by them nor the guides.

After the interaction, the tourists are led into the tiny City Walk Office, where they are told about the activities of the trust. Homemade posters on the walls show pictures of SBT-children who have grown up and attained a middle-class existence with an education, job and spouse, and one poster shows former guides, many of whom have gone abroad on scholarships. Pins stuck to a world map shows all the countries visitors and volunteers come from and a shelf holds awards given to the organisation along with the printed names of sponsors and aid agencies supporting it. The guides tell their personal story, hand out feedback forms and offer visitors to buy merchandise like pens and t-shirts with the name of SBT. The visitors then proceed to a separate office to make a mandatory payment of ₹200 ($3,5) plus whatever they wish to donate, and are then escorted back to the Main Bazaar from which they disperse.

How is this encounter with children whose fate is yet to be decided experienced by the tourists? A Danish high school teacher in his sixties whom I have called 'Jørgen' in this book described in an interview the transition from being on a 'tour' to interacting with the children in the shelter home. On the day he visited they were congregated on the roof instead of inside:

> When I am following [the guide] around, I feel very safe. Partly because he is there, partly because we are a few people together, partly because he wants to tell us something and we have made some sort of agreement. When we come up onto the roof [of the shelter home], then I enter into a private home, and then I have to be met in some way, so that I am not a bother to anyone. Because, I'm not 'there to see the animals', if you know what I mean. I am there to chat, or to hear about the organisation or to admire their work, or to understand the depth of the problems that are there in the society, whatever they might be. But there I get more insecure, because what do I wreck by coming there to watch?
>
> (Interview with 'Jørgen' 25:01–26:39, my translation)

As the format of the CW changes to be of a more intimate nature, Jørgen became anxious that his presence is detrimental somehow to the children, who neither speak English nor are of age, and Jørgen was therefore anxious that the 'agreement' he had made with the English-speaking CW-guides could not be extended to the shelter-home children.

How might Jørgen's anxiety be conceptualised? The classic distinction between fear and anxiety is that fear always takes an object, as you are always afraid *of* something. Thinking in spatial terms, Ahmed therefore defines fear as 'being produced by an object's approach', whereas anxiety is 'an approach to objects' (2004, 66), as it is about how the body approaches the world (nervously, hesitantly, anxiously?). Especially anxiety is thereby antithetical to the oblivious comfort of fitting within norms, as the anxious body is one that always questions the position it inhabits in space, and while Chapter 3 suggested that a small amount of discomfort might be titillating rather than unpleasant, it seems that Jørgen experienced this anxiety as unpleasant.

In the theoretical section of this chapter, the CW is conceptualised as a co-performed space of affective negotiation where im*press*ions circulate between the bodies as verbal and non-verbal communication in affective economies where signs gather affect in the circulation. Ahmed (2004, 70) suggests that affective economies of fear/anxiety might work via misreading, because signs of anxiety might be read as threatening and thus fear-inducing by a counterpart, who then, in return, acts threatening, thus creating a cycle where each im*press*ion moving between the bodies is invested with ever more fear. The same might arguably be true with anxiety, as the anxious approach of one body towards another might very well be met with a similar and perhaps amplified anxiety in the approached body.

One of the most important roles of the CW-guides is to put the CW-visitors at their ease so that the economy of affectively laden signs circulating between them doesn't enter into an escalating cycle of anxiety/fear, and instead enable the CW-visitors to move relaxed through the CW as if it were a comfortable space. At the beginning of the CW, the CW-visitors are therefore given instructions about how to act. They are told to be mindful of who and how they photograph and not to give money or food in packages to street children, because they might resell it and use the money to buy the solvents they sniff, and when the CW reaches the shelter home similar warnings are given about how to interact with the children. Apart from cautioning the CW-visitors and making them aware of dangers, these instructions also work towards stilling both fear pertaining to the harm that might be inflicted on them and anxiety pertaining to the harm they might inadvertently inflict on the children. This process of reassurance is part of the 'arrangement' that Jørgen refers to, but his exclamation '[*W*]*hat do I wreck by coming there to watch?*' in the preceding quote seems to indicate that he wasn't reassured.

Reflecting on the source of his anxiety, he continued: 'You know, I am a big man, and all things being equal they all know that I carry around an annual income and valuables and so on and so on. The distance is very, very big' (Interview with Jørgen 25.30–26.00 my translation). The economic disparity between them is simultaneously the reason the encounter has been initiated and the reason it is problematic. Linking economies of affect to economies of capital, we might thus see Jørgen's discomfort as originating from a 'fear of economy' and not just an 'economy of fear'. This is further illustrated as I asked him provokingly whether he was then engaging in 'poverty tourism' on the CW?

He stated that he in this context was probably a 'tourist' because he was travelling with his colleagues on a pre-planned round trip, but

> to be human is to try to understand, and I have no ... I would like to have the opportunity to talk to some of these people if possible, who live in ... who are also humans, who have hopes and wishes just like everybody, but who live in the proverbial cardboard box. And I cannot contain the fact that people live this way, I cannot understand that people can be that poor, so in that sense one could say ... No I think that 'poverty tourism' ... I do not dare call it that. That is derogatory in some way. I don't like that.
>
> (Ibid. 37:00–38:10)

To Jørgen, a respectful encounter between the global North and South, symbolised by himself and the man in the 'proverbial cardboard box', is possible due to their common humanity, but it made him uncomfortable to call attention to the fact that this encounter is inscribed within a discourse of tourism. Withdrawing emotionally from this exploration of the possible exploitation of the CW he concluded:

> The point is, I shouldn't feel like I am imposing, and they shouldn't feel that they are ogled as animals, I mean they don't, they are just some boys.
> (Interview with Jørgen 27:30–28:01)

Jørgen's pronouncement that they shouldn't feel like animals opens for the possibility that they can be viewed that way, which then leads him to dismiss the whole thing with the words *they don't* and to replace the category of *animal* altogether with *some boys*.

What we are witnessing is, in a sense, two affective economies of anxiety towards former street children. The first unfolds between bodies whose lack of linguistic abilities leads to the use of non-verbal communication, which increases the possibility of anxious misreadings circulating between the bodies. The second is a retrospective affective economy of anxiety about how to relate to 'the man in cardboard box', and even what to call the encounter – a prospect so daunting that Jørgen desisted from exploring it, even though I was the only one listening to this exploration apart from himself. This illustrates the interconnectedness between affect and discourse, as even the verbalised recollection and interpretation of an earlier non-verbal economy of anxiety might start the unpleasant circulation again.

Another reason this retrospective affective economy becomes anxious is due to the politics of naming. The stigma attached to signs such as 'slum' results in a sematic slippage described in Chapter 1 between connoting something depraved in itself or simply deprived of wealth, and thereby a slippage between something illegal or illegal*ised*. These connotations signify different explanatory models for how poverty and crime emerge, and the sign slum thereby comes to function linguistically as for example re-appropriated racial slurs, such as the n-word, which might legitimately be invoked by black people but not white. Even an articulation of solidarity using words with such multiple opposing connotations could thus be taken as an insult, and the economy of anxiety pertaining to of how these slippery words are interpreted doesn't stop until Jørgen stops explaining his actions and emotions as anything but an act of kindness towards bodies devoid of political markers of identity, that is '*some boys*'.

This reflects on how difficult it might be to voice dissent within the CW itself in front of the guides, who were at one point figuratively living in cardboard boxes. What is also striking is that he, in his eagerness to remove them from a victimising tourist gaze and instead see them as generic boys, also removes them from the socio-economic context that made him both initiate and question the encounter in the first place. In order to short-circuit the economy of anxiety and make the encounter respectful, he dismisses his 'fear of economy', which is to say his critical gaze on the effects of the economic disparity between them.

Reflecting on his urge to withdraw from affective exchanges of this sort, he concluded that some of the colleagues he was travelling with might be better equipped to interact with the children:

> Some of my female colleagues, among others the one you interviewed, are much better at things like this. She interacts with these kids much more immediately with wooden shoes on and gets something out of it, you know. I'm just not very good at that. I probably just sit down and talk to the supervisor.
>
> (28:39–29:52)

The reference to the 'wooden shoes' of his female colleague is a specifically Danish metaphor signifying a lack of sensitivity, which is viewed as productive here because it short-circuits economies of anxiety by making a conscious decision that everything is alright. In his urge to 'talk to the supervisor' (of the shelter home, Ibid.) he revealed that he'd rather know more about street life than meet the children who have experienced it, because he recognises that they don't actually possess the skills the communicate a context that might satisfy him intellectually. Instead of employing the 'guessing game' (Chapter 2) as a communicative strategy, he wished the guide would have simply engaged him in conversation, and he remarked that the group travelling with consisted of

> hardcore educators, who are used to talking, right? Who are used to asking. And if there had been a living dialogue, we would have probably squeezed him of every last thing he knew, and more, and what could then of course be the problem was that he found out that he didn't know that much.
>
> (12:10–12:45)

The CW must be an affective experience precisely because it cannot be an intellectual one, and so Jørgen recognised that the guides are selling the only product they have: emotions and authenticity. Jørgen wanted neither of them, and that is uncomfortable.

Conclusion

Chapter 4 began by asking whether critique is possible to articulate within the affective space of the CW, when all the feedback forms handed in after the CW are so overwhelmingly positive? What are the feeling rules that keep the affective space of the CW comfortable? By focusing on tourists who are tempted to stray outside this space of comfort, its boundary is outlined, and the analysis of an interview with a CW-visitor concludes that transgression of the boundary might be performed twice: on the CW and/or retrospectively in the interview afterwards, where uncomfortable, out-of-bounds-questions might be posed in confidentiality and relative privacy.

The methodological part of the chapter theorised a relation between affect and discourse, and this is illustrated when the interviewee experiences an anxiety in the interview similar to the one experienced on the CW. Part of his anxiety seems to be about the politics of naming, which is theorised in Chapter 1 as a semantic slippage inherent in the sign 'slum', which is also present in the sign 'street child', or even in what the interviewee chooses to call, 'the man in the cardboard box'. The slippage points to a struggle between different explanatory models for why poverty exists next to extreme wealth, and the analysis shows that this slippage makes invocations of solidarity with poor people contentious, and thereby potentially anxiety-inducing.

What exacerbates this anxiety is the interviewee's insecurity about how the process of remuneration functions between him and the street children he interacts with, because an act of solidarity might well be meaningless if no transfer of capital takes place alongside it. The theoretical section of the chapter conceptualises the CW as a space where economies of affect circulate parallel to economies of capital, where it is the job of the guide as an emotional labourer to ensure that the right kind of affect is circulating towards the visitors, so that remuneration might circulate back to SBT in the shape of a fee. On one hand, this takes place within a tourism economy, where SBT sells the experience of being allowed to co-perform the CW as a commodity to the visitors. On the other hand, it also takes place within a humanitarian logic, where visitors donate to SBT, and thereby to the current and former street children they care for.

In Chapter 3, the CW was shown to be divided into a performative track that highlights an authentic experience of Pahar Ganj as an informal, urban space, and a pedagogical tack that explains the long-term deprivation and abuse that street children endure. The former is characterised affectively by a sense of playful abjection, the latter by a sombreness befitting the suffering described, and they thereby might seem to enter into two different economies of capital. The performative track suggests that visitors pay a fee within an economy of tourism, while the pedagogical track suggests that visitors donate within a humanitarian economy. As the visitors reach the contact point and the shelter home, they encounter current street children whose futures are anything but certain. Suddenly, the visitors are next to bodies that in some ways constitute the space of playful abjection experienced within the performative track, but these bodies are also represented in the sombre, pedagogical track as suffering Others in need of help. The two tracks merge in this respect. But do the economies, with their differing logics of exchange, also merge? If so, how?

These are questions that lie outside the comfortable space of affective negotiation constituted by the CW, and as the analysis shows, transgressing the boundary between this space and what lies outside it is uncomfortable to the extent that the interviewee doesn't dare to do it on the CW or in the subsequent interview. The next chapter shows two even clearer examples of such discomfort, and it attempts to theorise the humanitarian logic applied on the CW.

Notes

1 The number of Danish tourists from 2001–2010 was 222,653 whereas the number of tourists from the UK in the same period was 6,266,931 according to Data.gov.in http://data.gov.in/dataset/number-foreign-tourists-india-2001-2010.
2 This is a trait also observed by Desforges (1998), in his study of how a 'travelled identity' translates into social capital for young travellers.
3 In this understanding of the interaction between body and mind there is perhaps really no difference between what might be termed 'physical' and 'psychological' as affect and emotions are as physical as anything else.
4 Drawing for instance on Butler's notion of *Excitable Speech* (Butler 1997) much of what Ahmed analyses in for example *The Cultural Politics of Emotions* (2004) is how the emotion evoked in written texts function in relation to the body.
5 Perhaps because the book is written before the 'affective turn' had taken hold.

5 The post-humanitarian logic of slum tourism

Soft- and hardcore poverty porn and ironic humanitarian appeals

The relation between tourism and poverty alleviation is contentious, not because tourism generally alleviates poverty but because it seems it could:

> In discussions on tourism and poverty, or tourism's role in addressing poverty, one important aspect of tourism is often not made explicit, namely that tourism conceptually stands in contrast to poverty. Poverty prevents people from partaking in tourism because other more essential needs have to be addressed first. Tourism is thus associated with wealth in the form of discretional income. I raise this point because it highlights the rather strange character of the idea that tourism can alleviate poverty. To ask whether tourism may alleviate poverty seems rather like asking whether wealth can alleviate poverty. This question is obviously misleading. Clearly poverty can be alleviated by wealth, but this is not what happens. To the extent that tourism is preconditioned on wealth, tourism can also alleviate poverty. But just as wealth does not alleviate poverty automatically, nor does tourism.
>
> (Frenzel 2016, 19–20)

As Frenzel correctly points out, tourism seems to be associated with wealth and this means that tourism is sometimes associated with poverty alleviation in rather naïve ways.[1] I would further argue that an association between tourism and displays of excessive wealth also means that slum tours, whose theme is poverty,[2] runs the risk of being exposed to outrage if their attempt at representing or alleviating poverty seems disingenuous.

One of the clearest examples of how outrage over the 'wrong' kind of slumming emerges and spreads is covered in Steinkrüger's (2016) study of the Emoya Luxury Hotel & Spa in Bloemfontein, South Africa, who constructed a 'shantytown' in a nearby game reserve with floor heating and Wi-Fi, as a getaway from the main hotel. Online reviews show that while few tourists reportedly saw it as an 'authentic' shanty experience, the slum theme was generally applauded as novel until the televised political satire show *The Colbert Report* made a feature on it calling it 'poverty porn'. The Internet soon responded with reviews on Tripadvisor.com, where some simply condemned it, while others posted spoof reviews pretending the Emoya shanty was part of an actual shantytown, thus highlighting the violence that its inhabitants might face if it were:

> The gun . . . gang battles outside the shack gave us a thrilling half hour, especially when a shot pierced the flimsy tin wall and lodged itself in my shoulder. This bit of the stay was only beaten by an unexpected morning wake-up call at 4am by the authorities accompanied by bulldozers, dodgy paperwork and an armed police escort who were there to demolish the 'shack' to make way for some commercial development. We had only a few minutes to grab our belongings and head for the door!
> My husband is still trying to shake off the TB infection he caught due to the insanitary and damp conditions and our kids are still blacking out over the psychological trauma and are refusing to leave the house, but at least we have a good story to tell at dinner parties!
>
> (Steinkrüger 2016, 244)

Compared to the performative and pedagogical track of the City Walk (CW), Emoya's performative track seems to have been sanitised to only include an exoticised version of poverty, which represents it as a simple life liberated from the choices of a modern existence, while the pedagogical track has been completely removed from the experience. It is only present in scathingly ironic reviews, like the one earlier, which enter into an affective economy of outrage as they are shared and commented on.

Analysing the relationship between the metaphor of *poverty porn* and what it describes, we might note that porn is a fetish of the skin.[3] Its affective logic is predicated on decontextualisation in that the messy business of ambivalent emotions between sexual partners, accidental pregnancies or sexually transmitted diseases is usually glossed over in its fantasies of casual sex. The forceful contextualisation provided by the earlier quote about the actual living conditions of shanty towns resembles the contextualisation one might provide to porn, by pointing to how sex and sex work takes place in real life, that is how the commodification of sex to create pornographic fantasies might sustain an exploitative economy. In both cases the pleasurable affective economy of exoticisation and arousal, respectively, is replaced by an affective economy of indignation and anger, which breeds more comments, reviews, blog posts and television segments, all confirming, sustaining and spreading outrage (rage-that-will-out) about this wrong 'slumming'.

But which representational practices then cause outrage in the global North and, more important, which don't? Frenzel's earlier quote might lead to the assumption that outrage over 'wrong' slumming has something to do with the proximity of wealth and poverty, but the target of *The Colbert Report* wasn't Emoya's main hotel, which like most luxury hotels signalled wealth and gave little or nothing back to the poor people surrounding it. It was rather the 'shanty' that portrayed poverty falsely that became the object of outrage. Similarly, it is not all interactions with poverty that might be susceptible to outrage. In an Indian context, many Rajasthani castles and fortresses have been converted into luxury hotels by Maharajas made landless by Indian independence, and stays here routinely include excursions into the small hamlets they used to rule. Often the inhabitants are much poorer than the average inhabitant of a slum in a large city, and more people die of conditions related to malnutrition and the proliferation of diseases that are comparatively easy and inexpensive to cure.[4]

Yet, while it is linguistically possible to go 'slumming', you cannot go 'ruralling' or 'villaging', and organised encounters with rural poverty then doesn't really have a name, even when such meetings are staged as re-enactments of historical power relations even older than the one between the urban elite and the proletariat described by Koven (2004) in a British setting. The reason for this could be that urban slums have historically been at the centre of political resistance to capitalism (Harvey 2012; Frenzel 2016, chap. 3), meaning that a discourse criticising and highlighting urban poverty has been more prevalent than one highlighting rural poverty, but the end result is still that rural poverty is naturalised and invisibilised to a larger degree than urban poverty in India, but even urban poverty in the global South seldom causes outrage in the global North, unless someone tries to portray it or facilitate an encounter between the rich and the poor.

It is also worth noting that the outraged review supplying a pedagogical track to Emoya's profile wasn't written as an earnest appeal to Emoya to portray urban poverty in a more nuanced fashion but, rather, employed an ironic, humorous register, which enabled its portrayal of violence to be consumed by like-minded readers as a *sublime* experience where 'the pleasure depends on being safe in this encounter with the overwhelming' (Dovey and King 2012, 12). The review itself might thus also be read as a kind of 'poverty porn', but the emotional reward for reading it doesn't come through Emoya's exoticised daydream of a carefree life but is derived from the pleasure of sharing righteous anger through an ironic stance, which crucially doesn't commit to an earnest declaration of solidarity with the bodies exposed to the violence portrayed.

Staying within the porn metaphor and employing the labels of the genre itself, Emoya's representational strategy might be called 'soft' poverty porn, as the pleasure lies in the ellipsis of sordid details, while the reviewer's might be called 'hardcore' poverty porn as the pleasure lies in the explicit inclusion of the same. But how did we come to a point where most all representations of poverty might be termed 'porn', and what outrages us is not poverty but how certain instances of poverty are represented or how certain interactions between rich and poor are staged.

In constructing a theoretical frame to answer this question, we are well served by the body of literature focusing on logics of humanitarianism and how these have been communicated over time as appeals to donate to organisations working towards the alleviation of the suffering of distant others. There is a long history of aid campaigns projecting spectacles of distant, suffering or vulnerable others in the global South into the living rooms of the global North, and these mediated encounters still I many instances preclude the physical interactions taking place in a touristic frame. They inform the discourses these physical interactions exist within, and influence how tourists from the global North relate to the pain of distant others, which suddenly becomes proximate in the physical encounters with urban subalterns that is the goal of slum tourism.

In her book on the shifting representational practices of aid campaigns, Lilie Chouliaraki (2012) claims that a shift has taken place in recent years towards what she calls a *post-humanitarian* approach to addressing publics, which is signified by a move away from highly emotional appeals to cosmopolitan solidarity within a 'theatre of pity', towards appeals that invite the public to engage in temporary, contingent and measured forms of emotional attachment within a theatre of irony. The move seems to be driven by more than 'compassion fatigue' or the 'I have seen this before' syndrome (2012, 60), where an overexposure to mediated atrocities leads to a desensitisation among publics that are being addressed so that appeals utilising grand emotions gradually lose their ability to affect these publics. It is perhaps also that the reason for helping is unclear, while the certainty that helping actually helps is undermined.

Chouliaraki draws on Luc Boltanskis (1999) study of the use of *Distant Suffering* in humanitarian media strategies, where he quotes Hannah Arendt's ([1963] 1990, 59–114) critique of what she calls *the politics of pity*, which stands in opposition to a *politics of justice*. Within the logic of the politics of justice, questions can be asked about how a group came to suffer, and this opens up for the possibility of an argument against helping, if they are seen to have caused their own suffering, while within the politics of pity such inquiries are seen as superfluous, if not obscene, as help within its logic should be given regardless of merit. But though the politics of pity has been efficacious through the 20th century to validate humanitarian aid campaigns precisely because it bypasses discussions of the cause of suffering, Arendt points out that the politics of pity thereby also cuts itself off from working towards a system that might prevent future suffering and contents itself with alleviating immediate suffering.

Boltanski elaborates on pity's inability to articulate a structural critique by explaining how it works as an emotion between concrete subjects, as opposed to an abstract logic that creates the foundation for a politics. Pity along with compassion plays a pivotal role in Christian humanitarianism and is found in for example the parable of the Good Samaritan. The parable illustrates that pity as an emotion depends on a distance between the sufferer and the person who feels pity, as between the Samaritan and the stranger he helped, in contrast with for example the parents of a suffering child, who are so close to the child that they would be seen as suffering *with* the child through a reaction of empathy[5] rather than pitying the child (1999, 3). As pity becomes the foundation of politics, however, this distance must be evoked between groups characterised by either good fortune or misfortune, and consequently '[t]he Politics of Pity regards the unfortunate en masse, even if . . . it is necessary to single out particular misfortunes from the mass in order to inspire pity' (1999, 4). The politics of pity thus homogenises the pitiable by creating a metonymic relationship between the individual sufferer, whom the fortunate should pity, and the group they should help in some way, and it simultaneously inserts a distance between a potentially helping, global public and the groups of unfortunate sufferers, whose cause for suffering cannot be structurally examined without straying into a politics of justice.

This has created what Boltanski calls a 'crisis of pity', which has affected media strategies in humanitarian aid campaigns, not only in terms of how distant sufferers are represented and positioned in relation to 'the public' but also in terms of how the public has been invited to act on suffering and how the subsequent delegitimisation of this helping public has changed over time. Chouliaraki (2010b) chronicles how early humanitarian campaigns play on the metonymic relationship between depicted individuals and the groups they are shown to represent, both in early 'shock effect appeals' and later 'positive image appeals', and drawing on Arendt ([1963] 1990), Boltanski (1999) and Höijer (2004), Chouliaraki exemplifies the former in her reading of an image originally captured in Northern India and used in a Red Cross humanitarian campaign in 1961:

> This image is a composition of people devoid of individualizing features – biological, such as their age and sex, or social, such as clothing. They are half-naked, exposing emaciated rib cases, arms and legs. Captured on camera, these body parts, passively sitting in a row as they are, become fetishized: they do not reflect real human bodies but curiosities of the flesh that mobilize a pornographic spectatorial imagination between disgust and desire.
>
> (2010b, 110)

Chouliaraki reads the image as situating the spectator ambivalently. If the image inspires the pity it is supposed to, it will homogenise the pitiable by inscribing a metonymic relationship between the dehumanised bodies and the mass of suffering others they represent. These others are distant, not only for the obvious reason that they are situated in India while the implicit spectator is not but also because the theatre of pity produces them as 'strangers' who should be helped regardless of their merit. They can be pitiable precisely because they are situated outside the society the spectator is a part of. But the image also employs a *colonial gaze*[6] on the racially and ethnically produced other, which the spectator is inscribed within as possible perpetrator, partly for historical reasons and partly because the spectator is reproducing this gaze by aligning him- or herself with the implied spectator. Within this logic, the spectator's production of the distant others in the image as pitiable might be read as a way of shirking responsibility for causing the suffering in the first place.

This is furthermore highlighted as the image enters into what Chouliaraki calls the inherent paradox of theatre. The communicative structure of the theatre is caught in an 'impossible duality; both [are] responsible for cultivating ethico-political dispositions in the polis and simultaneously culpable of producing narcissistic emotions' (Chouliaraki 2012, 30), and as we move from representations of fictional suffering in the theatre to representations of 'real' suffering in aid campaigns, the contentiousness of this duality is not diffused. The representation of distant others' suffering as real within aid campaigns is indispensable if it is to legitimise the appeal for aid. In the shock effect appeal analysed earlier, it is pivotal that the spectators believe that the photograph portrays the actual people they are invited to help, or at least that the suffering portrayed is real and will be alleviated, even if the bodies act as metonyms for a larger group of sufferers. But this realness might, in turn, serve to delegitimise its representational strategy, if the image is interpreted as giving the viewer a perverse pleasure, which is only heightened by the fact that it is represented as real, rather than fictional.[7]

In shock effect appeals, the spectator is thereby not only invited to align him- or herself with a colonial gaze that produces the spectator as shirking a responsibility, by receiving gratification for alleviating suffering caused partly by the spectator but also for experiencing the perverse pleasure of encountering 'curiosities of the flesh that mobilise a pornographic spectatorial imagination between disgust and desire' (Chouliaraki 2010b, 110). Compassion fatigue must therefore be understood as stemming not only from repeated exposure to distant suffering but also from the uncomfortably ambivalent position of the spectator.

As Chouliaraki continues her chronicling of humanitarian media forms, she argues that shock effect appeals were more or less replaced with 'positive image appeals' in the 1990s, exemplified by 'photos of smiling children, in the sentimental texts of child sponsorship or in the eyewitness accounts of aid workers' (2010b, 112). These are based on what Boltanski, drawing on Adam Smith (1793), calls a '*Sympathetic Equilibrium*' existing between sufferer and spectator, which is established by the absence of suffering, which has imaginatively already been alleviated by the spectator when the pictures are taken. The semi-pornographic spectacle of the suffering other is removed, but so is the suffering itself. The possible persecutor causing it is therefore also gone, as well as the implicit indictment that the spectator might be complicit in causing the portrayed pain. This is also problematic because

> [the images] fail, for example, to critically address the hegemony of neoliberal politics in world economy, the competitive governance milieu in which NGOs operate, the conditions of marketization and mediatization on which their legitimacy rests, the problematic links between NGOs and local regimes, as well as the lack of local infrastructures often leading to failures of development. In suppressing these complex dimensions of development, 'positive' appeals seem to lack a certain reflexivity as to the limits of the interventionist project to promote sustainable social change.
>
> (2010b, 113)

The 'hegemony of neoliberal politics in world economy' is not addressed nor any of the other ethical problems in the quote solved, even though the urge to help is felt as strongly as before, and because of this the earnest, emotional humanitarian appeals are transformed into post-humanitarian appeals, where cosmopolitan solidarity can only be expressed within a theatre of irony, as the earnest wish to help is undercut by the limits of the interventionist project.

Returning to the analysis of the reviewer on TripAdvisor who felt moved to supply a pedagogical track to Emoya's representation of a shanty, there are similarities between the charge of it being soft poverty porn, where the pleasure is predicated on the ellipsis of a messy reality and the charge against positive effect appeals recounted by Chouliaraki. There is however also what Chouliaraki calls a 'pornographic spectatorial imagination' in the review's representation of violence, but the anger in it is undercut by a humorous irony. Analysed within Chouliaraki's framework, that irony acts as a *mea culpa*, which acknowledges a possible complicity in not doing enough to resolve the underlying problems that cause the portrayed violence, as well as a complicity in turning the violence it berates into entertainment so that a discursively constituted public wants to engage with it, however briefly.

The similarity between slum tourism and humanitarian aid campaigns goes beyond modes of representation, however. The examples Chouliaraki gives of post-humanitarian aid campaigns ranges from marketing allegedly 'sustainable brands' promoted by celebrities (Richey and Ponte 2011), to attending major aid events like Live Aid and Live 8 (Chouliaraki 2012, 114–32) or to engage in online quizzes arranged by for example Action Aid focused on the individual donor's experience of giving rather than the receiver's need for aid (2012, 1–19), and part of what characterises these forms of engagement is that 'our moral encounter with human vulnerability is now cast in a particular logic of the market' (2012, 5).

She identifies two grand narratives of solidarity on which humanitarianism was formerly based: a Marxian 'solidarity of revolution' with oppressed peoples, who needed to liberate themselves from the capitalist system of oppression with the help of distant donors and sympathisers, and a 'solidarity as salvation' reaching back to Adam Smith, where charity was given to those who could not compete within the market. She writes the following about the transition from solidarity as salvation to *post-humanitarianism*:

> [W]hereas modern humanitarianism was grounded in the crucial separation between a public logic of economic utilitarianism, applicable in the sphere of commodity exchange, and a private logic of sentimental obligation towards vulnerable others, applicable in the sphere of individual altruism and increasingly in institutionalized philanthropy, late modern humanitarianism, what I here theorize as post-humanitarianism, increasingly blur the boundary between the two. In doing so it manages to turn the ever expanding realm of economic exchange into a realm of private emotion and self-expression and, in a dialectic move, to simultaneously commodify private emotion and philanthropic obligation.
>
> (2012, 5)

Post-humanitarianism is characterised by the perceived disappearance of the divide between the sphere of business and charity, just like contemporary,

108 *The post-humanitariansim logic*

globalised slum tourism, which is situated as both an aid performance and tourist performance. Just as consumers presumably alleviate suffering by consuming sustainable brands, slum tourists alleviate suffering by consuming an event and the line between 'commodity exchange' and 'obligations towards vulnerable others' are thus blurred in ways similar to what Chouliaraki describes.

Tourists going on slum tours are thereby positioned ambivalently in they are both consumers and donors, and this influences the subsequent moral judgement of whether they are truly alleviating the suffering they encounter, and if so, in which way? Parallel to this discussion is the question of whether slum tourism functions as soft or hardcore poverty porn, which is a charge that might be levelled against almost any representation of poverty. Emoya's complete lack of representation of poverty, as well as positive effect appeals, which draw on the thought of the *sympathetic equilibrium*, can be accused of glossing it over, while shock effect appeals, portraying it in all its horror, might be accused of wallowing in it. Even the angry-yet-ironic review of Emoya, which doesn't claim the moral high ground, might well be accused of entering into a self-sustaining, affective economy of outrage without providing solutions.

Given this analysis, the surprising finding of Chapter 4 is not that Jørgen is anxious about his position as a visitor but, rather, that most of the other visitors seemingly aren't. The next section further explores the feeling rules of the CW and analyses how it, as a space of affective negotiation, is kept comfortable by the

Figure 5.1 Interaction between tourists and SBT-children at Aasra Shelter Home

Figure 5.2 Tourists and SBT-children playing pat-a-cake at Aasra Shelter Home

emotional labour of the CW-guides, but also how certain visitors are compelled to transgress the boundary between this comfortable space and what lies beyond.

The anger of encountering shelter-home children

The analysis of Chapter 4 shows that Jørgen – a CW-visitor interviewed after the CW – has the impulse to explore what lies beyond the comfortable space of negotiation constituted by the CW. He has the possibility of doing so both on the CW itself and retrospectively during the interview but desists as the affective economy of anxiety becomes too unpleasant in both cases. The discomfort pertains partly to the naming of the street children – a dilemma that closely resembles the one theorised earlier – as well as a system of remuneration and what economies it enters into. The reason he desists is because he thinks it 'disrespectful in some way' (Interview with Jørgen), and his focus is thus on how he can avoid doing more damage than good.

A different reaction can be found in another Danish CW-visitor I have called 'Line', who was determined to see the post-humanitarian mixing of business and charity as a corrupt money-making scheme on the part of SBT. Responding to my first question about what she noticed on the CW, she said,

Line: There were two things. The first thing, that pops up at once, was the big box in the little office we ended up sitting in, marked 'Tips for the Guides'.

Tore: yes?

Line: that struck me, and then there was the meeting with the children that was arranged as little monkeys in a cage. (Interview with Line 7.05–7.30, my translation)

I then asked how she had perceived the interaction, and she said,

> I thought it was unpleasant because I they were 'Line'd up in rows, and when we come in, there was another group who had just left, so they had just acted as an exhibition for one group and shaken hands, and said 'hello sir' and 'mam', and played 'pat-a-cake' with jovial Americans, who loved sitting and doing stuff like that.
>
> (Ibid. 24.00–24.30)

Here, Jørgen's suspicion is fully formulated as an indictment that the children are being treated as an exhibition, but whereas he blames himself, Line blames the CW because she feels emotionally coerced into legitimising the performance once she has entered into it. Her refusal to take part in an interaction that validates a lacking system of care (as she sees it) might be read as a refusal to interact with the children, and she doesn't want that:

> [The children] were arranged/displayed in rows and you couldn't get out, my first thought was 'how the hell to do I get out again?' now that I had entered the rows, then I had to walk over them and go back, I felt very exposed, I thought 'this is very unpleasant', also when one came down to the bigger boys, whom I don't think were very charming, and I know I am being very direct when I say so, but I don't think they are.

Tore: It smells a bit of aggression?

Line: It smouldered! There were several of them who were getting into fights with each other at the end, where they had to separate them.

(Ibid. 25.20–6.10)

Here anxiety of accidentally offending them is transformed into fear of their possible retaliation, and the economy of anxiety/fear that Jørgen seeks to avoid is already a reality in Line's experience of the interaction.

At the time of the interview, Line had been staying in the suburb of Gurgaon with her husband and five children for two years as an expatriate. She accompanied another family visiting from home on the CW, and they consisted of a white couple in their fifties and their darker-skinned niece of nine, whom the couple had taken as a foster child. As they entered, the children raised their arms to the niece, who showed signs of being uncomfortable and quickly left with her mother. According to Line who later talked to the niece, she apparently compared them to the dead in *Hades* in the Disney film *Hercules* (Ibid. 50.00–51.00) reaching out to grab her, while Line herself described their movements as 'mechanical' and the children as having 'murder in their eyes' (Ibid. 27.12).

I read the children as being far less aggressive and certainly never outright dangerous, but I did notice that the physical resemblance between the children and the niece meant that the phrase 'there, but for the grace of God go I' had a slightly different meaning than it otherwise did in the relation between visitors and shelter-home children. The mysterious process of why some are born in wealth and others in poverty might be explained by God, coincidence or something else entirely, but in this case, it was the grace of adoption that separated the once orphaned niece from the fate of these other orphans, whom Line couldn't help but seeing as abject, scary others. And my sense from her narration of the incidence was that it was at least partly this uncanny closeness between the object of love and these other objects of abjection that was fearsome and thus anger-inducing.

As I asked Line why the format of the CW made her angry, she contextualised it with her own life as a comparatively rich expatriate in Delhi and linked it to what she perceived as the Americanisation of Delhi's social system that made her 'so F... [sound] because the problems are never solved' (Ibid. 22.20). She proudly claimed to scold her fellow female, American expatriates, who said that 'the poor were happy' and had 'glowing eyes' (Ibid. 29.50) and who did charity work while refusing to give their maids and drivers what Line thought was a proper salary and treatment. This fact made her suspect that most charitable NGOs were corrupt, and my claim that the organisation actually did help a lot of children didn't seem to satisfy her, either because she simply thought I was too dense to catch them cheating or because she thought the whole structure of lacking care that necessitated the organisation was corrupt, which meant that SBT also was. In any case, she told me that she was glad that the fee was only ₹200, as if the CW was a scam she had narrowly escaped.

Furthermore, the CW had not brought her any closer to the 'poor that scratches on my window' (Ibid. 12.18), and this was important to her, since her son had told her 'I don't see [the poor outside the windscreen] anymore' (Ibid. 12.40). She was thereby genuinely afraid that she and her family 'would become like the crowd' (Ibid. 42.20), which is to say de-sensitised to the poverty all around them, while periodically engaging in the kind of sensitisation that would demonstrate that charity, rather than reform, was needed in Indian society. Due to this fear, she felt inclined to transgress the border delineating the CW's comfortable space of affective negotiation at the moment where she felt that the dramaturgy of the CW prompted her to be overcome with positive emotions for the children. In the end she didn't do so in the interaction itself but ended up going through the motions of the encounter with the street children and listening to the personal story of the guide. In the subsequent interview, however, she seemed glad to be given the chance to turn her anger into outrage and transgress the boundary between comfortable and uncomfortable discursive (and thus affective) territory precisely because she felt it as the same boundary she was encouraged to keep within when moving in Delhi's expat circles.

What lay beyond this boundary was, as I understood it, the idea that the elite should give liveable wages for necessary labour performed by ordinary workers, rather than paying fees/donations to emotional labourers that staged performances

112 *The post-humanitariansim logic*

of their misfortune, which was then shown to be mitigated by the money paid by the tourists/donors to the NGO. Formulated within Chouliaraki's and Arendt's vocabulary, Line was equally disgusted with a politics of pity that bypassed discussions of why the poor were poor and a parsimonious politics of justice that accused the poor of causing their own poverty, because they seemed to work in tandem glossing over the fact that the poor were poor because the rich weren't paying them enough. This reasoning is not in itself opposed to the post-humanitarian blurring of charity and business, as the idea of for example fair-trade products is based on the workers receiving a fairer remuneration for their work than they otherwise would. But as Chouliaraki points out, this often leads attention away from the vulnerable and suffering others who need this type of aid and, instead, focuses on the donor's feeling while donating/consuming. How did Line's wish to replace charity with fair business dealing agree with her wish to encounter the poor in a meaningful way? Would the only way for her to meaningfully encountering them mean that she encountered them as something other than poor?

This paradox is similar to Jørgen's in Chapter 4. The economic disparity between him and the street children he encounters is simultaneously the reason the encounter is initiated and the reason it is problematic. And though Line's response is not one of anxiety but, rather, of anger, she also acknowledges that the poor street children can't be held responsible for engaging in a performance she dislikes, as it seems to be their last resort, and so she can't really express her anger at them. When she does so retrospectively in the interview, shame follows in the wake, thus exacerbating this anger in an affective economy where anger leads to shame and vice versa.

The grief and pain of encountering shelter-home children

Between Jørgen's anxiousness and Line's shameful outrage, we might ask how it is ethically possible to relate to the pain of others – especially if that pain is expressed in a performative format that you dislike because it is dictated by larger structures of inequality that you are complicit in sustaining? Or within the language of humanitarianism, how might the bodies of potentially helping spectators and suffering or vulnerable others relate to each other in proximate aid performances? Does the co-performed space of the CW allow for an ethical phenomenology of pain to exist between its participants?

This question became important when I encountered a 46-year-old Swedish woman I have called 'Agneta', who insisted on feeling pain on behalf of the shelter-home children and CW-guides, though they displayed no such pain in their performance of the CW. She insisted on grieving for them and what they might have been in a way that was both empathic and problematic, and this section proceeds to a quick theoretical detour to frame her response before proceeding to a description of it.

Chapter 4 theorised affective exchanges as leaving 'im*press*ions' on the surfaces of the bodies involved. These im*press*ions are really all we ever experience of other people, and while they are not identical to the bodies who left them, they might exist within us as presences we can interact with, even if their physical counterpart is not at that moment leaving its impression on us. And therefore,

> the parts of me that involve 'impressions' of you can never be reduced to the you-ness of you, but they are 'more' than just me'.
> (2004, 160)

This means that what we experience as 'other people' are a series of impressions left on us that 'live' within us, and so we live with the emotions of others – pain included – inside us, though these emotions are not identical to the ones experienced by the people they pertain to.

Ahmed (2004, 20–41) furthermore argues that pain is often externalised in some way, even when it – like menstruation cramps – comes from within. It is conceptualised as an interaction between the body and what is perceived to be outside it and might consequently be spoken about in metaphors such as a 'stabbing pain', invoking images of skin being penetrated.[8] But even though pain is often conceptualised as being caused by something or someone outside the body, it is never possible for someone else to feel the pain of another body. Ahmed therefore concludes that pain might be solitary but never private (2004, 29) because pain is interpreted as contingent, which is to say stemming from contact, though it is always produced by the body that feels it.

Empathy with the pained body of the other, like love, is a feeling that seeks to bring subjects closer to each other, while simultaneously reinstating the boundary between the two because the *desire* towards this transgression of the boundary separating bodies is a constant reminder that it is there. Empathy therefore remains a 'wish feeling'. Boltanski's (1999, 3) example earlier, where the parents of a suffering child would not be interpreted as feeling pity *for* the child but, rather, suffering *with* it is illustrative, because even if the parents do suffer when their child suffers, the suffering they feel is theorised by Ahmed as the impressions left in them by the child and thereby not identical to the suffering of the child itself. Proceeding to an 'ethics of pain' Ahmed (2004, 31) concludes that insofar as we have a duty to act upon others' pain, we have a duty to act on pain we do not know for ourselves, and we have a duty not to appropriate that pain while we act. We thus have a duty to keep the attachment 'contingent' (Ibid.) by not imaginatively collapsing the distance between the other that lives inside us and the *themness of them*. Drawing on Ahmed, we might thereby conceptualise poverty porn as a way of imaginatively appropriating the pain of the other, without feeling it as they feel it, which might result in a pleasurable but problematic, cathartic sadness.

Another way of relating to the pain and demise of the other is through grief. Ahmed conceptualises the grief over a death as a way of sustaining the relationship with the memory of a person that lives inside us, even if the physical counterpart is absent. Consequently, '[t]o grieve for others is to keep their impressions alive in the midst of their death' (Ahmed 2004, 160). Ahmed thereby departs from Freud's assertion that the diseased must be forgotten through the process of grief, if a damaging melancholia is to be kept at bay. She asserts instead that to 'realise' the loss of a person is to identify the impressions left on your body by the diseased, and to grieve not only for the dead body out there but also for the impressions inside you, which are no longer connected to that body.

But how do you grieve for the living and the pain they experience? When I encountered Agneta, she was travelling India independently for a month with her husband and two sons, who were 11 and 13. She was visibly moved by the visit to the shelter home, because while the other tourists interacted with the children, she sought to be alone and tried to hide that she was crying. I thought Agneta simply couldn't find her place in the interaction, as I had seen others not being able to, and that it might cheer her up to see some of the children doing something useful on their own, in a way where she didn't have to play such an active role. I therefore beckoned her towards a computer room in the shelter home, where three boys were sitting playing computer games, seemingly oblivious to our presence. Encountering these boys, however, she started crying again, and I gathered from this that whatever made her sad at that moment probably had less to do with the format of the interaction and more to do with the kind of life that the CW showed the children to live in the shelter home.

During the subsequent interview at a café, her sons and husband drank a cup of tea at another table, and as I asked why she had reacted so strongly to the CW, she started crying and said that the experience had brought home to her that 'it really happens'. By going on the CW she had been brought face-to-face with a reality that she almost couldn't bear to encounter. Her empathy with the children of SBT had surfaced as pain for them and made it 'real'. But to me, Agneta didn't seem like the traveller that would simply stumble upon the CW and be moved by it by chance, since the family was travelling independently and thus had chosen every experience on their trip. This meant that she might actually have sought this experience consciously, in spite of – or because of – the pain it inflicted. I therefore asked her about the pain and if she imagined 'that it might actually do some good that it is unpleasant?' to which she answered,

> Yes of course, I think so. If you get engaged or feel something, you are more inclined to do anything about it. If you have it far [*sic*] you forget it, and you don't do anything about it. So I think it must be painful, or . . . Yeah, I think it works better that way.
>
> (Interview with Agneta 1:10–2:00)

I asked her why travellers would go on such a tour on their holiday and she replied that people have a moral obligation to try to understand,

> because you see poor people. And you want to do anything about it. You can't just walk around and be a voyager all the time . . . it can't work like this [for] one month in India. It's not ok, it is your responsibility to learn more and to do anything about it.
>
> (Ibid. 2:30–2:53)

Incidentally, Agneta used the word *voyager* interchangeably with *voyeur* when talking outside the interview, and in the context of the interview the role that she described with the word *voyager* resembled a kind of unwitting voyeur who 'see[s] poor people' and doesn't even try to understand their lives and problems and help them in some small way. And so, Agneta felt it as a moral obligation to shed her role as this 'voyaging voyeur', who had neither the means nor the right to intervene, but precisely because the CW was set up as a two-hour tourist experience, she wasn't able to gain the knowledge that might have enabled her to do this, and this started the escalating cycle of anxiety/fear articulated in the words

> there are so many codes and so much we don't understand. And I feel like a stranger here,

and

> things you don't recognise, it's often scaring [sic.]. It's just the fact that you don't know why it is like that, and that's . . . maybe that is uncomfortable.

However, there was also

> that feeling of guilt. Why are they having it like this, 'cause I . . . I don't think anyone wants to live like that. And I can't feel, are they angry with me? They should be angry. It's so unfair. Why do they have to live like this, when we got money? It's so unfair. And that's unpleasant.
>
> (Ibid. 6:48–8:10)

So, Agneta was anxious because she was aware of her ignorance of the context she was in, and consequently she could not be sure that her reading of the locals and CW-guides or children at the shelter home was correct or to what extent they felt resentment towards her privileges, a resentment that she herself felt while simultaneously taking advantage of that privilege.

Contrasting this reaction to Jørgen's reaction, Agneta bore no hope that decontextualising the encounter and simply seeing the children as 'some boys' might short circuit the economy of anxiety/fear, because she saw the disparity between them and her as essentially 'unfair'. On the other hand, she didn't feel Line's urge to make an angry transgression of the borders delineating the comfortable space of affective negotiation on the CW either, precisely because she didn't feel entitled to do so. Her response was therefore one of grief.

Referring to the smiling demeanour of the CW-guides, I asked her if she thought that the CW-visitors perhaps sometimes had a stronger emotional response to the telling of the CW-guides' life stories than the CW-guides themselves and why that was. She reflected that their response would probably have been stronger if it was the first time they told it and that though she believed it to be terrible to lose your parents at an early age, like this particular CW-guide, his way of diminishing the emotions connected to the retelling of his suffering 'must be good because people have to adjust to survive. People can get used to almost everything' (Ibid. 9:40–10:10). She believed that if it happened to someone in 'our little town in Sweden, it would feel more unfair' (Ibid. 10:20–10:30), but she didn't want to copy this relativism because 'I can realise that it is like that, but I can't, no it doesn't comfort me'. I asked, 'Would you like it to?' and crying again she said,

> No. I want to feel it! And I want to remember it. I never want to forget it. Because then I might buy less things for myself. Give more money instead, in the future. I hope so.
>
> (Ibid. 10:30–11:26)

As I asked her about how she imagined her children perceived the CW, it turned out that her moral obligation was also theirs. She told me they actually didn't want to go on the tour, 'but we said, you have to do this, this is the most important thing we do in this month in India' (Ibid. 2:53–2:59), and in terms of the long-term effects she imagined this might have, she said,

> *Agneta:* It will come, questions in one week or in one month, and we will talk a lot about it. Right now it's just hard for them to see, I think they do like this (shielding her eyes) a bit they can't take it in.
> *Tore:* They keep a distance?
> *Agneta:* Jah jah, and . . . still they understand, I think they understand that it is serious, that it is reality. . . . we'll talk about it a lot and they'll have many questions, and I saw that they were very serious, and they can't understand very much, and we're telling them about . . ., so I think they have many questions.
>
> (Ibid. 3:04–4:28)

To Agneta, her level of sensitisation was at least partly a pedagogical project directed at her children. She expects the CW to create an emotional need in them to ask questions afterwards, which corresponds to her opinion that unpleasantness is good, when it comes to educating oneself in how to respond to the misfortune of others. However, Agneta was the only one on that particular CW who showed any other emotion than a polite reflection of the CW-guide's smiling optimism and the only show of apparent concern I observed in her children was towards their mother during the interview, where she cried openly while talking to me. After the interview they came over to our table, and she, her face bloated from crying, smiled at them, while I, feeling the urge to signal that I wasn't the source of their mother's grief, did the same, thus indicating that the interlude of sensitisation was over and normality was restored.

Conclusion

Comparing the three interviews in Chapters 4 and 5, it seems that the discomfort identified in all cases stems from similar ethical problems, though the reactions to this discomfort are quite different. Jørgen in Chapter 4 fundamentally believes in the possibility of a respectful encounter initiated by a socio-economic divide between its participants, but he is anxious that the minors he engages with on the CW cannot refuse to participate in it. He furthermore recognises that the CW commodifies poverty to some extent, and he is uncomfortable with how that might situate him as a 'poverty tourist', but perhaps even more so with how this might lead to misplaced assumptions of victimisation of the shelter home children on his part. In order to short-circuit an affective economy where his anxiety about approaching them might be read as fear or dislike of them, he refrains from mounting any critique at all, both on the CW and in the subsequent interview.

Line is deeply uncomfortable playing her part in the interaction with the shelter-home children or to let herself feel the relief at the end of the CW-guides' personal story. She does not accept these performances of 'hardship overcome' as something that will change the lives of the SBT-children for the better or as the cure for the de-sensitisation she and her family is experiencing while staying in Delhi, and as she experiences the fear of her niece as her own, the encounter for her turns from being merely a disingenuous performance that substitutes real engagement with emotion, to being downright dangerous, thus igniting her anger. This leads her to level allegations of corruption against SBT, though this is only substantiated by the post-humanitarian practice of mixing business and charity. The very concept of slum tourism is based on this mixing, and it seems she might support it in other instances but not here. With no actual proof of corruption and no competing system of charity to point to, Line cannot vent her anger at SBT and even less at the boys that represent the NGO, and this lack of a way out seems to enrage her as much as the CW itself.

Agneta, on the other hand, acknowledges that the tragedy she sees in the lives of the shelter home children and CW-guides is not felt by them, but she sees this as a mechanism of repression that they cannot afford to discard given their circumstances, and this to her is all the more heartbreaking. To Agneta, she and her children thereby have a moral obligation to feel the grief the shelter-home children and guides cannot afford to feel, because she, as opposed to Line, believes that this emotional involvement will translate into socio-economic change. Agneta therefore attempts to hide her feelings from the shelter home children and guides, while her own children are meant to feel it with such intensity that she projects her feelings onto them before she has even talked to them about it.

The question is, though, what does she then grieve over, phenomenologically speaking? Using Ahmed's framework, we might conclude that she's grieving over the impressions in her of what the guides could have become, had they had the chance. Simultaneously, she recognises that they cannot indulge in the luxury of grieving in the way she thinks they ought to, and she therefore not only grieves *over* what she imagines their lives to have been; she also grieves *for* them, that is *instead* of them. But the im*press*ions left in her that she is grieving over/for has not been left by the guides in the first place. Ahmed's ethics of pain, which dictates that you don't appropriate the pain of the other's body while interacting with the impression of that pain, is in Agneta translated into an ethics of grieving

118 *The post-humanitariansim logic*

for the guides' who can't grieve, thus imaginatively instating a loss in them that she can grieve for.

But just like there is a difference between the tourists' interactions with the topos of Pahar Ganj experienced in the performative track of the CW and the actual hardship of the street portrayed in the pedagogical track (described in Chapter 3), there is also a difference between what Agneta insists on grieving for and what the guides grieve for. They, of course, have lingering emotional attachments to past im*press*ions of parents, siblings, friends and volunteers, which they sustain or terminate in their own processes of grief. Agneta doesn't insist on involving them in her grief for them, and she thereby doesn't seek to appropriate their process, but as the next chapter shows, emotional outbursts still affect the guides, whether they are meant to or not. It therefore contains an analysis of how they work towards protecting themselves from such outbursts in their performance of the personal story on the CW.

Notes

1 How Frenzel conceptualises the generation of value through tourism is covered in Chapter 4.
2 See Chapter 3.
3 Porn can be seen as yet another illustration of Ahmed's point that im*press*ions circulating between bodies exist both as non-conscious affect and as emotions existing in the symbolic order, because while pornographic images of tabooed skin and penetrated bodies might arouse other bodies gazing at them at an unconscious, affective level, what is seen as taboo varies between cultures and is thus negotiated within the symbolic order.
4 Writers such as Saninath (1996) and Gupta (1997) point this out.
5 Though as Sara Ahmed (2004) reminds us, they would not feel the actual suffering of the child but rather the 'wish-feeling' of empathy – a wish to feel the suffering of the other, to take it away or to share it – see also the section on Affective Approaches to Tourism Research.
6 Chouliaraki connects the image to Stuart Hall's (2001) concept of the colonial gaze and thereby to the tradition of Postcolonial- and Subaltern Studies examined earlier. While she is not explicitly concerned with Subaltern Studies, there are similarities between the genealogy that she (by way of Arendt) and Chatterjee draw on, in that both Arendt and Chatterjee critique enlightenment ideals from a position on the margins of a European society that claimed authorship of them (for a description of how Chatterjee does this, see Chapter 1). Arendt's argument in *The Origins of Totalitarianism* (1958) is that rights given by membership of a polity is lost when membership is lost, and she bases her analysis on a pre–World War II Europe, where fascism and Nazism strips Jews, communists, handicapped and homosexuals of their rights by creating them as the other of the nation state, as well as a post–World War II Europe, where streams of refugees leaving war-torn countries like Germany are stripped of their rights in turn. And as we learn from *On Revolution* (Arendt 1990 (1963)), what prevents them from receiving a similar treatment is really only the politics of pity, which circumvents discussions of merit. As Chouliaraki reads images of suffering Indian bodies mediated to a public situated in the global North as pornographic, it seems we are once again back to the question of how subaltern population groups can represent themselves covered by Chatterjee.
7 The rise of reality TV in the 2000s illustrates this point. Familiar narratives of love, sex and betrayal are enacted by supposedly 'real people', who are prompted by the 'reality

shows' they appear on to enact what was previously fiction. Simultaneously, the proliferation of sci-fi movies projecting a dystopic future where reality TV has commodified all aspects of human existence, like *The Truman Show* or *The Hunger Games* shows that the very 'reality' that some find desirable is also what might be thought of as reprehensible (see also Boltanski 1999, 12).
8 Hurtful words are also often conceptualised as 'wounding', which again shows that words circulating between bodies are not just bearers of abstract meaning divorced from emotions but, rather, are conveyers of immediate sensations that are tied to the production of meaning.

6 The emotional labour of CW-guides

Collecting data on the CW-guides

My attempts at getting to know the CW-guides was by far the hardest part of my data collection. The difficulty I had with obtaining a version of the CW-script described in Chapter 2 was an indicator of the trouble I had in getting them to open up, and by the end of my stay I was much closer to a group of younger SBT-boys that I taught music, along with a pair of retired guides I stayed with, than with the guides I was meant to work with. Chapter 7 describes how that came to be and how I conceptualise my position within SBT and the exchange of different forms of capital between me as a researcher and them as interlocutors. This section focuses on how I developed and collected different kinds of data about the CW-guides.

As I started following the CW and recorded what took place on it, I came to learn that the different guides really all performed the same CW, except for their personal story, which was told at the end of the CW, right before the visitors were asked to fill out the feedback form, pay the mandatory fee and the CW was over. I therefore started to collect and compare these stories, while seeking to gain knowledge about what had perhaps been excluded from them. I spoke to the guides, who were reluctant to talk, and to SBT-staff who were slightly more helpful. After three-and-a-half months of my going on CWs, teaching them English afterwards and generally hanging out, most of the guides perceived talking to me as work. Some liked it because they learned things; others thought it boring or too demanding, while one particular guide, whom I have called 'Rishan' in this book, demanded that I buy him cookies and refused to take me seriously unless I 'bribed' him in this way.

With one month to go, I started setting the scene for more formalised ways of extracting data. I elicited the help of SBT-City Walk & Volunteer Coordinator Poonam Sharma, who wanted some sort of formalised evaluation she could show her superiors, and this gave me the mandate to ask the guides to sit down and talk to me about their jobs and lives. This had the unfortunate side effect that they thought that whatever they told me would then reach the ears of their immediate boss, though I, of course, insisted that the answers were confidential.

Happily, most of what they didn't want her to know consisted of tactics to avoid doing tasks they disliked or learning things they didn't have the patience for, and most of the outright lies or omissions on their part seemed to take place in relation to this topic. One of the things I was interested to learn was how they policed themselves and their peers to appear to be good SBT-children and CW-guides, and so even if they did exaggerate the amount and importance of the work they did for SBT, this projection of what they wanted me to think was useful.

During my stay, I had become friends with a long-term volunteer called Nick, who was about my age, and a former CW-guide called Danish, who ran SBTs volunteer flat where we all stayed along with the changing population of volunteers to the organisation. They had known many of the guides from when they first arrived at SBT, and they invited them for a meal at the flat cooked by me, along with the former SBT-City Walk & Volunteer Coordinator Jessie Hodges, who was American but still settled in Delhi. She, along with Nick and Danish, staged a conversation with them about certain topics that I perhaps couldn't really breach with them on my own, and this directed me in what questions to ask them in the formalised interviews, along with the personal stories I had recorded. As it happened, the food and the drink also earned me some respect within the ranks of the guides, and one Assamese guide even hugged me after the meal, because he missed the mutton curry of his home state and my version of the dish made him feel at home after many years in Delhi.

Thus prepared, I conducted the interviews, presented the overall findings to Poonam Sharma as an evaluation of my stay and returned to Europe to analyse my material and apply for further funding. I got it, and the following winter I found myself back at SBT for a follow-up session, which revealed other aspects of the narrative structure of the personal stories, as well as the emotions they might elicit. The CW now had its two first female trainee guides, who at that time hadn't yet internalised the unwritten rules about how to narrate a personal story, and the girls' initial attempts at forging one revealed some of the hidden rules the male guides' stories followed, precisely because the girls broke them, and were given suggestions about how to 'improve' them by the older, male guides in feedback sessions that I recorded. Second, while the overall structure of the older guides' stories didn't seem to have changed much, it struck me that the most experienced of them had attained enough linguistic and narrative proficiency to shape the stories to suit how they wanted to relate to the audience of the day. They could now more or less control the 'mood' of the room while narrating, and this made another analysis possible of how they gained control over the emotional labour they performed.

I thus arrived a data set rich enough to outline how the CW-guides navigated between some of their narratively constituted identities, and though I could, of course, have continued my fieldwork and explored other aspects of their lives, it would have taken another book to do it justice.

Figure 6.1 A CW-guide explains the poster of 'success stories' among former SBT-children at Aasra shelter home at the of the City Walk

The shaping of a guide's 'personal story'

Analysing the life stories of the CW-guides, Margaret Wetherell's (2007) concept of a 'personal order' is efficacious in theorising how stories shape our identity. Within her framework, the continual narration of our lives never stops and takes place in the most unlikely of settings, from whispered confidences to loved ones to banal explanations of why we are late for work. The real and imagined events of our lives are connected to each other by narrative logics of causality, teleology or coincidence in a web that functions as an explanation of who we are, in the ever-moving 'now' that constitutes the time of telling. A 'personal order' might be visualised as the growing 'backbone' of a subject's life story that becomes stronger with each iteration, and, while it orders events into sequences, it also imbues them with meaning that informs the identification process of the narrator.

The concept of the 'personal order' is theoretically located within Wetherell's later writings, where she moves from an 'ethno-methodological tradition' and instead approaches a 'genealogical' tradition, where the former is characterised by somewhat decontextualised conversation and discourse analyses framed by social psychology (e.g. Potter and Wetherell 1987), whereas the latter takes into consideration the wider social context that subjects articulate their differing subject positions within, partly in terms of the interpersonal relationships and histories the subjects enter into and partly in terms of the wider, traceable genealogy of knowledge production that the narrative draws on in a Foucauldian sense.[1] Linking this to the phenomenological framework of this book, we might utilise Sara Ahmed's conceptualisation of bodies as 'congealed histories of past approaches' (Ahmed 2004, 160). Within each 'approach', affective and discursive im*press*ions move between bodies, which means that what we experience shape our bodies and psyche, in a process where the two finally cannot be conceptually separated from each other. Combining this with Wetherell, the theory of the personal order might function as a framework for analysing how such 'past approaches' are imbued with meaning through stories, which have their own logics separate from that of the body.

Wetherell emphasises that the different positions taken in the performance of a life story often conflict internally, and that all subjects thereby perform the role of 'dilemmaticians' (Wetherell and Potter 1992, 198), who operate in response to a series of conflicting frameworks, where the narration of their stories work towards achieving certain goals. A 'personal order' is thereby 'personal' because it is felt as such, not because it is forged in isolation from social contexts, and Potter and Wetherell (1987) stresses that each iteration of a life story is told *to* someone, who might at any time intervene, and the narrator must thereby necessarily occupy a discursively constituted subject position in the telling, which a listener is at least willing to interact with, though the position's validity might be contested.

Relating this to the wider theoretical scope of this book of how CW-guides are able to represent themselves, Chatterjee (2004, 2013a) claims that though 'political society' provides a space from which the contemporary, urban subalterns (or 'the governed') might speak, they are also thereby forced to frame their identities within the discourse of governmentality that deals in groups rather than individuals, in order to get access to services they rely on to survive. The subaltern are thereby theoretically able to speak (Spivak 1988), but if subalterns want to be heard and understood they have to speak a language, and within a discourse, understood by the elite, and if they want political influence they have to speak of themselves as a constituency, and thereby a group or collective body. Based on this, Chapter 2 showed that CW-guides are positioned as *liminal subalterns*, perpetually on the verge of complete resocialisation and thereby able to simultaneously represent current street children and SBT, as an organisation that resocialises them.

This position is also articulated very clearly in the personal stories of CW-guides, but the former SBT-City Walk & Volunteer Coordinator Jessie Hodges, shows in her unpublished MA thesis from SOAS[2] (Hodges 2011) that this liminal subalternity has a much longer history of articulation. Based on fieldwork conducted in 2010, she finds that young children who come into contact with SBT start shaping their stories in ways that will make them eligible for help, and consequently they start referring to themselves as 'street children' because this is the category of children that the organisation has a mandate to work with. There is thus a circularity in the categorisation of the 'street child', because the discourse available to children who seek to inhabit this category produces them *as* 'street children'.[3] Second, there is a homogenising trend in the narration of the 'street child's' story, as little space is left to the narration of events that situates them in subject positions that ambiguously linger outside that category.

The process of shaping the 'personal story' of a future CW-guide thereby starts early and is right from the outset not only 'social' in the sense that the shaping is performed within a space of negotiation but is also part of a struggle to get access to the services SBT offers. Trainee guides are usually admitted to the CW-program at the age of about 15 or 16, at which time they would have been asked to tell their story many times to SBT-staff, in the beginning to evaluate whether it would be feasible to 'rehabilitate' them to their parents, to ascertain the amount of psychological damage their experiences have caused and to help them build a narrative identity that might hold these experiences in a way where the possible trauma attached to them is gradually lessened as time goes by.

As the trainees join the CW-program, they are presented with three goals to be achieved: (1) learn to speak and write English, (2) learn the CW-script by heart and (3) construct a 'personal story' about themselves to be told on the CW. In practice, the volunteer-teachers attached to SBT help the trainees achieve these goals within the same didactic frame. English proficiency is taught via the example of the CW-script, and the CW-script, in turn, holds general information about the lives of street children. Together these give a language and a context for the trainees' initial attempts at a personal story to be narrated at the end of the CW. At the same time, they play the role of co-guides on the CW, where they follow more experienced guides around and listen to their performances of the script and story. This provides inspiration for the narrative format of the trainee's personal story, and if it hasn't already done so, an 'order' – personal but constructed within the didactic frame of the English lessons and the CW – will emerge with each iteration.

Analysing the personal stories of seven male guides in 2013 and two former male guides, a series of patterns emerge, the most prevalent being that the ending is always the same. The end of a story, understood in narratological terms (Brooks 1992; Sarbin 1986), is the point in the plot where the original conflict of a story is resolved. In the guides' personal stories, this point is reached when the guides have become 'success stories' like the young men and women on the poster in the SBT office, and joining SBT is therefore an event that is always included in the stories as a pivotal moment of redemption, because that is the act that allows them to concentrate on this journey towards success, whereas the ending of the personal story is not, in fact, situated at the time of telling but, rather, in a teleologically constructed future where goals formulated in the present have been achieved.

Performing such an ending, however, is something that is learned, and an illustration of this fact was performed during my second period of fieldwork in 2014, when I observed how one female trainee's personal story in the making fell short on this account. When trainees have created a preliminary version of their personal story, it is customary that they tell it in front of the other guides and the CW-coordinator to get feedback, and in one such session I was allowed to tape the proceedings on video. On the recording, the telling is performed quite fluently, and in the feedback session afterwards everyone is appreciative. After a few comments I asked what the most important thing in her story is, and pausing as if to search for the correct answer she restated her aim with the exact words that she used in the story: 'My dream. To become professionally skilled so that I can support my mother' (Durga's personal story).

Juxtapose this to the level of detail in the performance of even a newly trained guide obtained the year before:

> Now I complete my high school and I complete my graduation from Delhi University of commerce, and after I have graduated maybe I will [inaudible] or get a job in tourism industry.
>
> (Kiaan's personal story)

Similarly, the ending of a more senior guide's story is constructed thus:

> So presently I am studying in High school through distance education and I have done a six-months course in multimedia and basic computer. I joined this program one year ago because I wanted to improve my English communication and gain confidence. And I also want to join travel and tourist industry in the future.
>
> (Jai's personal story)

When the trainee states that her dream is to become 'professionally skilled' and nothing more, she is able to point to herself as an object of SBT's pedagogy, but she cannot yet state precisely how this process will save her. The other guides can enumerate their past, present and future degrees and careers, and as they do so, they perform their own future internalisation of the discourse of personal achievement mediated through a personal story, which becomes a success story in the making. It would, of course, be more convincing if they were able to actually speak within the discourses they are trying to internalise in fluent English, but the basis of the CW as a touristic and humanitarian performance is that the guides are able to speak for the urban subalterns trapped within Delhi's informal sectors, which were once their peers. In the approximately six years they are a part of the CW-program, the guides must thus represent themselves as being in a perpetual state of *becoming* resocialised, because if the process finishes while they are employed as guides and they stop referring to the subject position of 'former street child' as one they are trying to escape, then they will not be able to convincingly represent the street children they claim to speak for. Seen in this perspective, it is only natural that the ending of the personal story is situated after the time of telling, since that is the time when complete resocialisation is supposed to have taken place and not before.

The conflicts established in the beginning of the personal stories are much more diverse than their resolutions. What they consist of depends to a large degree on what has in fact transpired in the narrator's life, and since the guides' lives have been quite different from each other before they encountered SBT, this is perhaps unsurprising. However, there are similarities because the first part of any personal story always explains how the narrator came to be in a position where SBT could play its redemptive part, and hence the conflict needs to be constructed as something that can indeed be resolved by that act. This opens up for the necessity of '*plotting*' (Brooks 1992), and as the guides become increasingly adept at this, their personal stories begin to be *about* something more than themselves and hold more than a mere description of the suffering they have personally endured and how they were saved from it.

They thus 'theme' the conflicts of their personal stories, and Hanuman's story is one example of this. It starts in Mumbai, where he is separated from his parents and sent 1,700 km north to the city of Patna, where he is enrolled at a children's home, which he escapes from. The story then concentrates on ways to survive on the street until he gives up that life and joins SBT. Here he is talking about the 'fake drinking water trick':

> I saw it happened in station that some street children were filling water bottle from the tap and selling inside train station, sometimes the kind of water you tourists get sick from. We also filled our own bottles from a tap . . .
>
> (Hanuman's personal story)

This, however, does not provide a secure life, so when he is contacted by a gang he at first refuses:

> But they started telling me, 'if you live with us you can not find the problem from the police issue and local public'. I started living with them and joined the gang and I learned how to pick pocketing. I was the youngest pickpocketer in that group. Like eight and a half. Slowly, slowly I became a professional pickpocketer living on the street.
>
> (Ibid.)

The gang then goes to Delhi to 'earn more money', but though he encounters several social workers who encourages him to join them, the idea that apparently moves him to join SBT and leave the gang is the dream of having a steady job as a train conductor, though this dream is later supplanted with dreams of being in the tourism business. The title of the story could thus be 'Resocialisation: From Scallywag to Budding Citizen'. The narrator's shifting affiliations illustrates the impossibility of surviving by yourself on the street for any length of time, and encountering SBT offers him a way out of this life.

As a contrast we might turn to the personal story of Ali, who, like four of the seven current guides, decides to run away from a poor, abusive, dysfunctional home:

> [M]y father used to play gambling a lot. He lost all our money in gambling and his addiction made us a poor family and caused trouble almost every night at home. So one day I did a small mistake and for that I was beaten very badly. That time I realized: why should I live here? He always trouble my whole family and he always beat me.
>
> Then I planned to run away from home. Like one night, I took out 3,000 rupies out of my father's pocket. In morning time, I wore my school dress, put some clothes into my school bag and I told my parents that I'm going to my school, but I instead took a bus for the railway station and I came to Delhi by train. Like that time I was about nine and a half years old.
>
> (Ali's story)

Later, Ali's story also focuses on shifting affiliations as a means of survival and the punitive measures of the police, but unlike in Hanuman's story the conflict here is constituted by the loss of a home, and redemption can therefore only be achieved by finding a new one, and this 'search for home' is another very common theme in the personal stories.

In stories with this theme, a certain 'point of no return' can be traced from the earliest personal stories told by guides in 2006, which I found in SBT's digital archive, to four of the seven personal stories told in 2013. It is a sequence of events that starts with a description of living in a family with abuse followed by the theft of an amount of money and the escape via some mode of transport, usually a train. Including the theft in the narrative serves a double purpose. It explains how it is practically possible for the narrator to leave, like Hanuman's story of selling fake drinking water or picking pockets, Ali's theft explains children's acts of petty crime as their way of surviving in an adult world. But it is also inserted as an act of dramatic betrayal that would make it hard for the narrator to return to his family without suffering a severe dose of the treatment that drove him away in the first place. It is a figurative bridge burning in the background, as the young protagonist makes his way towards the Indian megapolis and an uncertain future, and as such is an compelling motive found in many stories where children – mostly boys – strike out on their own to meet their fortune, from Dickens's *Oliver Twist* (1841) to Adiga's *White Tiger* (2008). As we shall see, it is also something of a simplification.

A third group of personal stories includes ones where the guides never actually live on the street before they encounter SBT. In the personal story of Jai, the narrator's elder brother originally runs away and comes to live with SBT, but returns as an adolescent to his village and convinces his parents to let him take his younger brother (the narrator) to live at SBT. Because options are limited and food is scarce in the village, his parents agree, and the narrator thus travels directly from his home to SBT. Similarly, the guide who hugged me for cooking him mutton tells a story of how an SBT social worker contacts him half an hour after arriving at New Delhi Railway Station from his village in Assam.

To sum up, the similarities in the narrative structures of the personal stories indicate that trainees borrow themes, details, plotlines and endings from the older guides and thereby contribute to the continual production of a narrative group identity as 'former street children', which then provides them with a position they might inhabit in relation to both SBT and the CW-visitors. The analysis also shows that the endings of the personal stories are situated somewhere in a not-too-distant, unequivocally happy future, which is facilitated by SBT, and the encounter with the SBT social workers is therefore represented as a pivotal moment of redemption for the guides.

Excluded stories and ironic performances

Some of the narrative elements not included in the official personal stories were passed on to me in private moments by SBT-staff, long-term volunteers like Nick Thompson, former guides and the current guides themselves. As I collected and ordered them into alternative narratives, they opened up for the possibility of other plotlines, themes and endings, which situates the guides differently in relation to both SBT and CW-visitors, while also complicating the relation between the guides and the street children they claim to represent. One striking feature of these alternative versions is that they complicate the narrators' compartmentalisation of subject positions and the narration of the transitions between them as concrete and irrevocable.

The narrative model for how the guides stage their escape with its 'point of no return' is challenged by the fact that most of the guides, and indeed most street children, run away in stages. Some have 'false starts', where the transportation they thought would take them to the city only brings them some of the way or in the wrong direction. Others are caught by authorities or members of their family before they reach Delhi, and even if they reach Delhi most children are 'rehabilitated'[4] to their respective homes, while only a minority remains 'runaways', largely by refusing to give up the names and addresses of their families. One clear indicator of this is that many of these guides, who initially position themselves as orphans in their personal stories, contact their parents and siblings once their position within SBT is secure, and they trust that they won't be sent them back to live with them in the countryside. The prevailing tendency among these cases is that the former guides, who have established themselves in one position or other, help their families and siblings, rather than the other way around, and so Danish, a retired guide who ran the volunteer flat where I stayed during my fieldwork, had helped his brother with a lawyer's fee when the brother was arrested for fraud, while Iftekar, whose story is told in Chapter 7, helped his brother start a business selling handbags in Delhi. Recognising this tendency, SBT has made it their unofficial policy to not insist that children they encounter must be sent home to families they do not want to live with, to the extent that they even allowed one guide to migrate voluntarily from his family to join his brother at SBT.

Just as the actual relationships between families are marked by a more gradual detachment and reattachment than the personal stories suggest, the redemptive meeting with SBT social workers and the happy ending situated in a not too distant future are also challenged. As Jessie Hodges (2011) explains, street children generally build complex networks of support to enhance their self-reliance, and these include other street children and adult neighbours, who provide food and clothing and religious sites, where food and shelter are sometimes provided. It even includes family, though this does not always mean that the children return in any permanent sense, and so just like they run away in stages, they are also resocialised in stages:

> Seen from this perspective, organisations are in fact part of street life rather than a way out of it: they count as one coping strategy among many street children access.
>
> (Hodges 2011, 15).

This, however, is an unwelcome fact for many of SBT's donors, and especially the Indian Council of Child Welfare, which has over the years taken progressively punitive measures against non-governmental organisations (NGOs) working with street children, in order to ensure that the children stay within the confines of their shelter homes in order to uphold a clear spatial distinction between those formally in need of help and everyone else.

This distinction is undermined by the fact that most of the children run from poverty and hunger rather than individualised abuse, and even if individualised abuse occurs in other personal stories, the frequent mentioning of poverty and hunger means that it cannot be separated from the socio-economic context it exists within. The 250,000 farmer suicides in the 1990s and 2000s (Sainath 1996, 2011) as well as India's 60 million malnourished children (Gragnolati et al. 2005, xiv) are poignant examples that vast areas of rural India suffer from conditions of deprivation usually found in disaster or war zones. This means that the guides are a part of a much larger group of children, most of whom are not helped, and whose problems are too big to be solved by NGOs such as SBT. The circularity of children labelling themselves 'street children' and telling a 'street child's story' in order to get access to services offered them by organisations such as SBT is thus initiated by children navigating in this field, but while it serves their needs it also validates the NGOs working with street children because it delineates these children as a separate group that might meaningfully be helped. And as we shall see, this separation also helps the CW-visitors connect emotionally to the children they meet within SBT, as opposed to the children they encounter outside, whom they are told they cannot help by randomly donating to them.

There are also events that are excluded because they describe shameful and/or illegal activities, though they might otherwise fit into the format of the personal story of the street child, and, if anything, exacerbate the position of the narrating guide as either an authentically criminal street child or alternately a deserving receiver of the visitors' donations. They include abuse of a sexual nature and crimes committed that, unlike the 'drinking water trick', don't fit the figure of the Dickensian scallywag, such as stories of serious gang fights or muggings, where opponents or victims have been left in uncertain circumstances, which might result in jail time for the narrator, if it is ever publicly known who perpetrated the violence.

These stories of violence inflicted or borne were generally told I private moments, whereas others were merely whispered, like the story from a retired guide about the mass rape of a runaway girl, which he may or may not have participated in, while living on the railway station as a young child. He confided in me one late night after a couple of beers, and before I went to bed I wrote down the story in my field notes, remembering his finishing sentence: 'She had to be everybody's girlfriend until she was the girlfriend of one. There were about 100 boys living on the station' (Field notes). The story was meant as an illustration of why girls running away from home at an early age could sometimes be better off enduring the dull, structured violence of the brothels on Gastan Bastan Road just north of the railway station, rather than the unpredictable violence of the street, but it also points to the prevalence and normalisation of sexual abuse, which males expose each other to on the street. Many of the young boys that would later grow into CW guides thus endured relationships of dependency, where they were systematically raped by adults or older boys, but I only learned this from case files and conversations with Poonam Sharma and Jessie Hodges, not from the guides themselves.

All in all, the biggest difference between the personal stories presented to tourists and the private stories told in other settings seems to be that early the betrayals and suffering included in the private stories cannot be reduced to plot-functions such as minor impediments on the way towards a 'happy future'. Partly because of the trauma attached to these events, partly because it is sometimes hard to believe that a happy future, in fact, awaits. And even if it does, there is always the question of what happens to the guides' families, especially the siblings, who are often trapped in the disastrous socio-economic contexts the guides have escaped.

Listening to the staged dinner conversation with Jessie Hodges and Nick Thompson, I found that the superficiality of the personal stories suited the guides, and as I interviewed them separately, they confirmed this. Ali, whose personal story included the theft of his parents' money and the escape to Delhi by train, explained the exclusion of certain events this way:

> Talking about when we were on the street and how we struggle, talking about struggling life, what we really face. But we can't say everything what happened to us. And people [the guides] they don't feel happy to say everything. Like suppose there is happening something wrong with you, you can't say that thing to everyone. You know something secret also happened, so we can give some hint, not explaining everything. We can give them some point. If they are clever they can know the point, they can understand, just get that.

I don't think so it is required, you know, and guide will also not feel good to say everything, because you know, maybe they can be very emotionally – again they have to come again in depression and think it about that, so that is not good for the guide as well.

(Interview with Ali 56:40–57:00 [SIC])

The interview started with Ali showing me a new tattoo on his wrist that inexpertly covered a scar from a wound. From its position, it might have been self-inflicted, but he merely said it was from his time on the street and asked me what I thought. I told him was nice. It was a green hippie symbol from the parlours of Pahar Ganj catering to backpackers looking for a memory of their trip to India, and while it drew attention to the scar, rather than covering it up, I thought it perhaps served the same function as the personal story, a sign to strangers that they should leave the scar underneath alone.

Another guide, who had been rescued a few years previously from a sweatshop, where he had been forced to sew purses, explained the truncated version of his personal story this way: 'I only get sad when [the visitors] get sad' (interview with Kabir 11:20–11:25), thus stressing the importance of keeping the mood in the room light. In a similar manner, Rishaan, who had previously demanded cookies from me, joked that he had other strategies to deal with visitors who become emotional:

We say 'don't worry, we'll give you a tissue paper'...

(interview with Rishaan 9:00–9:10)

Rishan not only states that he has tissues at hand for when someone starts crying; he also mockingly calms the hypothetical crying visitors in his statement by implying that a lack of tissues might be their biggest problem, perhaps because they get to continue to the next tourist attraction after the CW, while he must live the life they find tragic.

As I returned in 2014, it seemed that especially Rishan, who was now in his final year on the CW, had used his added English proficiency to perfect this ironic stance towards his own story and the visitors listening to it. After three years as a trainee and almost three years as a guide, he had grown tired of it and had begun to perceive telling it as a job he might as well make interesting, lest he got bored. I recorded his performance, which begins with a guessing game:

Rishaan: So friends, now I am going to tell you about myself, are you interested?
CW-visitors: Yes!
Rishaan: What is my name?
CW-visitors: Rishaan!

Remembering someone's name is crucial if a space is to be established where confidences might be comfortably exchanged, but since the names of the guides are generally unfamiliar to most of the visitors, they tend to be forgotten between the first introduction and the narration of the personal story an hour forty-five minutes later. The quiz is a cheeky reminder of the touristy frame of the CW, and communicates to potentially emotional visitors in the group that excessive emotions will not be welcome. He continues:

Rishaan: Good, so my name is Rishaan, when I was five years old, I left my home because my father used to beat me a lot and my mother always support my father. One day my family get separate. The consequence was: me and my other brother take with my father and my small sister when with my mother. I stay with my father for a few days, then he realise that I can't earn money for him, so he left me at the market.
CW-visitor: uh!
Rishaan: I was very [noise on recording] but he didn't come back again. While I was waiting for him, I met a couple of people who told me, 'I am your aunty and uncle' and your father will never come back again. Just because of that I ran [to] their home and started working for them. They always beat me, and sometime they put chili in my eyes.
CW-visitor 1: *They put chili in your eyes? [sounds incredulous]*
Rishaan: Afterwards I thought, 'if this is life I don't want this life anymore'. I planned to run from there . . .

The story is one of betrayal, abandonment, deceit, violence and, finally, desperation, prompting the narrator to escape to an uncertain future at the age of five. Yet it is told in a matter of fact way, with pauses that leave space for the visitors to interject exclamations, and the story proceeds with a comical reference to the fact that the visitors might get 'Delhi Belly' (diarrhoea) if they tried eating the discarded food he was forced to eat while surviving on his own. Though a horrible detail, it receives laughs all round, and soon he is back with yet another quiz, this time about which movies he likes, Hollywood or Bollywood? They guess wrong, as his favourite movie is the Hollywood production *The Pursuit of Happyness* (2006). And true to that statement, the story has a happy ending teleologically situated like all personal stories in the future, and the CW-visitors suspend disbelief in this happy end with these departing words:

Rishaan: So is there any more question you have about my life?
CW-visitor 2: *Just that we admire your . . . but your English is quite good actually.*
CW-visitor 1: We admire your guts.

As visitor 2 cannot decide whether to compliment the guide on his guts or his English proficiency, he ends up doing both clumsily, though visitor 1 finishes the statement. The theme of the story could thus be 'guts' because the suffering of the narrator is teleologically inserted into the story as impediments on the way, that developed his character, and the triumph of overcoming this suffering overshadows the grief that it had to happen at all.

The guides' long process of constructing a personal order, by narrating their lives in relation to external frameworks that might ensure their survival, thus ends with a professionalised version of their personal story. It might be flippantly ironic, like Rishaan's; contain euphemisms, like Ali's; or simply be quiet like Kabir's, but they all exclude certain events. Even Rishaan's inclusion of horrific events overshadows the sexual abuse I was told he had endured, and so all stories seem to have a residue of events too shameful or hurtful to be included.

Second, they all serve the function of controlling the mood in the room where they are told. In Chapter 4, the guides were theorised as emotional labourers upholding the balance between economies affect and capital so that these economies might continue to be mutually constitutive in that the right kind of affect circulates towards the visitors while capital circulates the other way. The preceding analysis shows that the shaping of the personal stories seeks to limit the guides' emotional involvement, and when Ali says, 'I don't think so it is

Figure 6.2 A CW-guide expands on the merits of former CW-guides at Aasra shelter home at the of the City Walk

required', about revealing excluded events in the earlier quote, he discursively produces the telling of the personal story as a job, which requires a certain level of emotional labour, though not total honesty. The boundary separating the comfortable space of affective negotiation on the CW from an uncomfortable one, which the interviewees in Chapters 4 and 5 transgress retrospectively in order to explore their sense of discomfort, is thereby constituted not only by the format of the CW but also by the amount and type of emotional labour the guides are willing to engage in.

Subaltern shame and performative therapy

According to an extensive *Guide Training Manual* compiled by Jessie Hodges when she was CW-coordinator, the CW has a number of goals, one of which is '[t]o sensitize people about the reality of street life and the hardships faced while on the street' (Guide Training Manual, Hodges 2010, 49). In an interview I conducted with the CW-coordinator Poonam Sharma towards the end of my stay, I asked what this meant, and she replied,

> [M]ost of the people who actually see street children [think they] are like street dogs. They don't go to them. I mean, how often have you seen actually people going to street children and saying 'where are you from?'. 'Are you lost?'. 'Can we take you somewhere?'. 'Can we take you to police station, can we give you . . . ' No! So, for them they don't exist.
> (. . .)
> 'Sensitizing' means that they should not ignore the fact that there is one group of people, who are there because of specific reasons and that they need help and that help should be provided to them. They should not be ignored. They should not be made in the margin of society. Street children are not street dogs.
> (Interview with CW-coordinator Poonam Sharma tape a 42:50–43:20 & tape b 2:55–3:20)

To Sharma, the sensitisation process that the visitors have to go through is supposed to transform street children from what Goffman (2008, 84) would call '*non-persons*' into visible, deserving receivers of aid, but what she is perhaps thereby overlooking is that visitors from the global North go willingly.

Theorised within the touristic logic of money for commodity, the visitors actually pay a fee to be sensitised, while theorised within a humanitarian logic, they are showing that street children are deserving receivers of aid by donating to them. Similarly, all visitors interviewed for this project seek a respectful encounter with urban subalterns on the CW, and the three interviewees represented in Chapters 4 and 5 experience discomfort precisely because they feel that they perhaps don't do enough in various ways. So, rather than forcing visitors to relate to urban subalterns, the CW perhaps provides a space that feels so safe that the CW-visitors pre-existing wish to be able to help in a meaningful way might be fulfilled. A space where they dare to be re-sensitised for a while and engage in the emotionally satisfying act of giving, before they are once again left with the overwhelming poverty of Delhi, where they are forced to de-sensitise themselves and turn away street children for their own good when they ask for money, because, as the CW-script claims, 'they will use this money in wrong way' (CW-script).

It is therefore important for the affective logic of the CW that the children within this space of re-sensitisation are separated from those outside it, and the personal stories fit this logic perfectly, because the sudden and irrevocable changes of subject positions in the narratives as well as the teleologically constructed happy endings uphold the idea that the guides and shelter home children may be saved, as opposed to those situated outside this space, while the guides' excluded stories undermine this assertion, because they illustrate that they are still part of the same social context that produced their initial problems and that they are perhaps not really solved, though they do live better lives because of SBT.

To be fair, the people Poonam is referring to, who must be forcibly sensitised to the plight of street children might be Indians, who see street children every day in Delhi, but only 3% of the visitors are Indian according to the feedback forms. If they did go on the CW, they might be enlightened, but there is also a good chance that they would relate to the personal stories like they relate to the guessing game represented in Chapter 2: politely staying within the boundary of the comfortable space of affective negotiation, though they are aware of the complications not included in the explanation.

The personal story at the end of the CW links the image of a lost, urban subaltern who engages in the kind of reckless behaviour described earlier, to the current liminal subaltern standing in front of the CW-visitors in the shape of a responsible CW-guide. According to the story, he will continue to improve himself beyond the time of telling, until he reaches the story's happy ending and thus complete resocialisation. The visitors are positioned as facilitators of this process by transferring capital to SBT in the shape of a fee/donation but also simply by being present and talking to the guides. The guides state on the CW that it was started partly to 'train [the guides'] English proficiency and presentation skills' (CW-script), and so the CW-guides finish it with the words 'thank you for giving me the opportunity to practise my English' (Ibid.). The CW is thereby framed as an educational experience, which the CW-visitors are, in fact, facilitating.

Second, telling the personal story in a public forum is communicated as a vital part of the process of resocialisation, not only because the guides improve their English proficiency but also because the teleologically constructed happy endings help them commit to a plan for the future. The mere act of attending a CW and listening to this story is thereby communicated to the CW-visitors as contributing to the guides' pedagogical progress both linguistically and psychologically, and the visitors are thereby positioned as not only as donors and consumers but also as benign interlocutors invested in this process.

When asked about the guides' psychological progress, Poonam Sharma reflected on the possible therapeutic potential of constructing and reiterating such as a personal story:

> There was a great author that said: 'If you can tell your story, you can become anything. So, if you can share your story, which means you have accepted your thing and you're ready to move on . . . so that is also one of the reasons why we do this. Because then it really has a therapeutic effect, unlike if you don't share your past and you keep brooding over it and you feel embarrassed, you feel that you are inferior and all that, so it tends to show somewhere or the other, whereas if you share it, you get a larger platform, which means that you have accepted it, you are ok with it and you are ready to move on.
>
> (Poonam 13:50–15:00)

To Sharma, it seems the personal story not only helps the street children envision a future but also work towards getting rid of the shame of having been a street child and to grieve over the suffering inflicted in some sort of healing process. But since the exact parts of the guides' lives that might need emotional processing are excluded from the official, personal story, what therapeutic potential does it have?

Frenzel (2016) makes the observation that poor people living in slums and the tourists who come to visit them are usually connected in a shared shame. He traces this shame back to the creation of 'the poor' as a social category by western intellectuals such as Rousseau, which gave rise to the question of guilt, because if poverty is not a natural state it must be continually created *by* someone, either the poor themselves or an elite who allow poverty to exist. In this there is an echo from the argument in Chapter 1 between Wright and Engels about whether the slum is populated by depraved people or whether they are simply deprived of the wealth that would allow them to live better lives. This then leads to the show/shield debate of whether to represent the poor and, if so, how.

Frenzel asks,

> If slum residents are feeling ashamed, then who told them to be and with what intention? If local elites insinuate that slums are shameful, is it the right answer for tourists to just ignore their existence?
>
> (Frenzel 2016, 8)

Are local elites hiding their own shame or the poor's when they argue that their presence should be ignored? Who projects which kind of shame onto the slum and street children? How do they do it? And how does shame then work within street children when they grow up? A short theoretical framing of question is necessary at this point before returning to an analysis of this specific case.

Sara Ahmed (2004, chap. 5) describes shame as painful, because it involves being exposed, while trying to cover up, a burning, reddening sensation of the skin that urges the body to turn in on itself while it is being turned towards others.

There is pain in shame, but it cannot be ascribed directly to others, as no one is inflicting it on the body apart from itself. Similarly, it resembles a reaction of self-disgust, as it involves the urge to pull away from something that is part of the body, but performing disgust is an act of separation that frees the body from what is deemed disgusting, while the shameful cannot simply be pulled away from. It sticks to the body and keeps reminding it of itself and how others are witnessing it. Shame is also different from guilt in that it refers to the quality of the person being shamed, whereas guilt refers to an action perpetrated by that person. This results in a radical difference in how guilt and shame are ascribed to others, as a person might be found guilty by others while pleading 'not guilty', while to shame someone means trying to make someone feel a sense of shame. Similarly, you might be guilty without feeling bad, while you cannot be shamed without at some level consenting to it; you must know and feel you have done wrong to feel shame.

What is curious, however, is that shame might also be felt over actions not perpetrated by the shameful, which is why sexual crimes committed against the guides might well be felt as shameful even though the guides are blameless in this respect. Feeling shame is a way of reaffirming an ideal or norm you haven't lived up to, so to feel the shame of inflicting pain on someone else is to reaffirm the importance of not doing so. But ideals and norms you do not agree with at an ideological level can still shame you, so that inhabiting an abnormal body – in terms of the shape it takes in the world, the signs it displays, the other bodies it desires or, indeed, how it has been unwillingly penetrated by other bodies – can still inflict shame, even if its owner doesn't think it should. Understood within Ahmed's framework, our bodies are congealed histories of past approaches, and because of this, norms shape the image of the body we feel we ought to have, and the 'wrongness' some bodies feel when they rub against spaces that haven't taken their shape is often a shame that feels all the more unpleasant because it feels oppressive.

The poor Indian body who is not able to perform the role of consumer for a lack of funds in an Indian megacity devoted to consuming might thus burn all the more with shame by a sense of perceived unfairness directed at the unequal distribution of wealth. Similarly, a body that has been abused by members of an adult world, which, moreover, tends to blame the children who possess this body for the abuse, might well burn with that same mixture of shame and perceived unfairness. The guides manage this shame by constructing their personal stories so that they only include mildly shameful events and crucially only things that more or less all the guides have been exposed to. They thereby discursively and affectively create the subject position of 'former street child' as less shameful than what it might otherwise have been and situate themselves squarely within it.

Using the theoretical framework outlined in Chapter 3, their shaping and performance of personal stories can be analysed as creating the CW as a comfortable space. A space they don't rub against if they stay within its boundaries, by only performing the agreed-on version of their identity. Consequently, when visitors try to transgress the boundary of the CW's comfortable space of affective negotiation, the guides don't welcome it, and this is so obvious that even Agneta, who insists on grieving on their behalf in front of her own children, feels that the guides have no use for her grief and tries to hide it while on the CW. Conceptualising the guides as emotional labourers, one might then see the simplified features of their personal stories as limits to the emotional involvement they are willing to enter into in order to sensitise the visitors, and as long as the visitors either believe the stories, or recognise the limit to the emotional involvement they are supposed to elicit, the interaction on the CW remains a comfortable one.

Conclusion

Chapters 4 and 5 showed how hard it is for CW-visitors to transgress the boundary delimiting the comfortable space of negotiation co-performed on the CW, and Chapter 6 continued this analysis by showing how CW-guides perform the emotional labour of shepherding CW-visitors into it. Outside this space, the visitors dare not be sensitised, lest it is revealed that the money and engagement they are prepared to provide are either too small to make a difference or even do harm. Within this space, however, they are told that their limited donations and engagement are appreciated and make enough of a difference, and consequently, the CW-visitors might allow themselves to be temporarily re-sensitised to a suffering that they are now seen to alleviate.

The personal stories performed at the end of the CW are instrumental in delineating the boundaries separating the group of children that can be meaningfully helped by the visitors from those who can't. The stories are characterised by a compartmentalisation of a series of successive subject positions, which always ends with the teleologically constructed happy ending situated in a not too distant future, and this edits out the description of how embedded the guides still are in the socio-economic context they are trying to escape, both as networks they rely on to survive and in terms of their lingering attachments to im*press*ions of friends and family they left behind when they joined SBT.

By including the events and emotions they only utter in private moments the guides could construct stories that were less callously happy, but they have no desire to do so, because the emotions these stories would elicit in the visitors might affect the guides emotionally, while it might also undermine the work of SBT and thus the position of the CW-visitors. Consequently, CW-guides increasingly learn to demarcate the limit of the amount and types of emotions they are willing to accept from the visitors, by shaping and performing stories that don't invite such transgressions of the boundary surrounding the space of comfortable affective negotiation or only do so ironically.

The next chapter focuses on volunteers, who involve themselves to a larger extent and donate more than CW-visitors. Does their added commitment translate into a different kind of sensitisation, and how do SBT-children and -staff react to it?

Notes

1 I derive this characterisation of Wetherell's shift in methodological approach from Benwell and Stokoe (2010), who provides an overview of the tradition of discursive psychology.

2 *The School of Oriental and African Studies* is a part of University of London.
3 This points to a larger trend of what for example Dip Kapoor (2005, 2007) calls NGOisation, where the behaviour Hodges points to are repeated in the relation between NGOs and the bodies that fund them. So, while the individual subjects seeking help from NGOs shape their personal stories to fit within its scope, the NGOs also shape their programs according which funds they might access by doing so, leading to a system in which 'The Tail Wags the Dog' (Kapoor 2005, 211), in the sense that what counts as problems to be alleviated are determined from the top down, rather than from the bottom up.
4 This was the word used within SBT to signify that children were reunited with their parents, whether they wanted to or not.

7 The economy of resocialisation
The slumming researcher?

Scripts of involvement and detachment in volunteering

When I first presented Salaam Baalak Trust's (SBT's) Volunteer & City Walk (CW)-Coordinator Poonam Sharma with the idea that I conduct my fieldwork within the organisation in the summer of 2012, she suggested to me that I did so by working as a volunteer, and I agreed, thinking that this might give me access to the former street children living within the organisation. At the time, I had an MA in English and had recently written two guidebooks on India for a Danish publisher, and it was therefore agreed that I should teach English and 'presentation skills' to a group of CW-guides, aged 15 to 21. From 10 a.m. until noon they generally conducted their CWs, and in the afternoon they studied via distance education programs, helped by the ever-changing population of volunteers, which came to include me. Because I had worked as a musician and a music teacher in Denmark, I was also charged with teaching music to a group of children, ages 6 to 18, who lived in an SBT shelter home donated by the Delhi Metro Rail Corporation, called the 'DMRC home'.

Arriving in Delhi in the winter of 2013 for four and a half months of fieldwork, I was accommodated in a 'volunteer flat' owned by SBT and run by a long-term volunteer called Nick, who was English and, like me, in his late thirties, along with his protégé whom I've called 'Hakim', who was a former CW-guide but now divided his time between studying at Delhi University and running the flat when Nick was in England working as a notary. Hakim's friend 'Iftekar' dropped by from time to time to visit us, and he often interacted with the transient population of volunteers living there, many of whom were girls his own age. One such volunteer was 'Beate', who stayed for a month and a half in the flat and worked at the DMRC home.

My relationship with this group became pivotal for how I came to understand the CW-visitors, CW-guides, long- and short-term volunteers, SBT-staff and their interplay. My daily conversations were recollected each night in my field notes, which I periodically consulted when developing my mode of inquiry in the studies presented in this book so far. They furthermore came to constitute a set of empirical data in themselves that provides insight into questions which lay beyond the scope of these studies. How do former CW-guides navigate in relation to SBT after they 'graduate' from the CW? It was beyond the scope of my budget to simply stay for three more years to find out, so instead, I examined the social setting of the retired guides at the flat. How might CW-visitors perform in relation to SBT, if they extend their involvement beyond the two-hour scope of the CW? During my fieldwork, none of them did, but I found that examining the reactions of the volunteers might provide insight into this question.

Second, my interactions while 'hanging out' at the volunteer flat furthermore enabled an analysis of how I as a researcher was also part of an affective economy. Chapters 4 and 5 showed that such an economy exists between CW-guides and CW-visitors and that the circulation of affect and capital have to balance on the CW in order for it to exist as a comfortable space of affective negotiation. My field notes from my time at the volunteer flat show, however, that other kinds of capital were also circulated in economies existing between myself, the volunteers, CW-guides, other SBT children and the SBT-staff, and this chapter ends with an analysis of how I also came to perform, both off- and on-stage, as a facilitator of this economy and how I gained from it in certain ways. First however, we will turn to an analysis of the emotional scripts volunteers act in relation to while volunteering.

In her study of the emotional responses to encountering poverty among volunteers in Kenya, Emilie Crossley concludes that

> tourists . . . are demanding impoverished places as their holiday destinations. In this sense, poverty tourism has affinities with the more established practice of volunteer tourism; the only real divergence between the two being the level of tourist involvement in trying to reduce poverty.
> (Crossley 2012a, 235)

In Crossley's study, 'involvement' thus seems to separate volunteering from slum and poverty tourism, and from the quote it seems evident that more involvement is valued higher than less, meaning that voluntourism is valued higher than slum tourism among volunteers because of its level of involvement.

Judging from for example Vodopivec and Jaffe's (2011, 119) study of volunteering practices in a Latin American context, being 'involved' doesn't mean that prospective volunteers go where their specific skills are needed, and in the marketing of VSAs (Volunteer Sending Agencies), emotional attachment and the willingness to work seems to be the only qualifications young volunteers need. Consequently, '[m]aking a difference is advertised as something that can be done by anyone with good intentions who is willing to work hard' (Vodopivec and Jaffe 2011, 119). And even though some of the volunteers in the study find out that specific skills are needed, such as Spanish-language skills when teaching Spanish-speaking children, this apparently hasn't seemed self-evident before they went.

This somewhat naïve idea of how to alleviate the problems of the global South might well be linked to this book's genealogy of how the colonising nations have historically represented the nations they colonised (Said 1978, 1993; Bhabha

1994; Spivak 1988), which is continued by writers like Ilan Kapoor (2004). He shows how this power-relation is repeated in the very idea of what constitutes 'developed nations', as opposed to 'developing nations', and thus how the conception of 'development' perpetuates a series of unhelpful dichotomies, even as it is used to point out the unfair distribution of capital globally. I would suggest this as a historical context for, for example, Simpson's (2004, 668) finding that the volunteers in her study has a tendency to locate poverty outside their own societies, making it a feature of the less 'developed' societies they are visiting, an observation that also resonates with Crossley's findings that a type of 'spatial Othering'[1] (2012a, 236) of the host countries take place in the discourse of the volunteers.

Vodopivec and Jaffe, however, also notes that while their interlocutors among the volunteers often do regret their lack of relevant skills while volunteering, very few take the step further and voices a disenchantment with the neo-liberal discourse that theorises the suffering they encounter as caused by individualised misfortune, rather than embedded in larger structures of inequality that can only be dismantled via collective, political action.

In this vein, they continue:

> [M]any members of the current generation of young people are caught between, on the one hand, this sense of personal responsibility to make a change and, on the other, the structural difficulties of doing so. This frustration is compounded by the growing awareness that corporate actors are poised to commodify any desires, hopes and ideals they may cherish.
> (Vodopivec and Jaffe 2011, 121)

Vodopovic and Jaffe thereby identify a reluctance to fully commit to an emotional engagement in voluntourism that is similar to the post-humanitarian logics of engagement conceptualised in Chapter 5 of this book (Chouliaraki 2012). Appeals invite publics to engage in temporary, contingent and measured forms of emotional attachment to causes, where logics of altruism and business are deliberately blurred, and given the temporariness of these forms of attachments, Chouliaraki sees these appeals as often being performed within a theatre of irony, rather than the theatre of pity where such appeals, following Arendt, were previously made. However, Chapters 4, 5 and 6 of this book showed that this ironic engagement is reconfigured when participants of the CW attempts to create a space of comfortable affective negotiation, where this proximate aid performance might take place. An ironic attachment to past suffering is performed, but mainly by the guides, who use it to distance themselves from a harmful level of emotional engagement, which they might otherwise be exposed to when performing their personal story three times a week in front of a group of strangers.

Crossley's (2012b) longitudinal study of how volunteers understand themselves as morally changed by encountering poverty during their time abroad is helpful in contextualising the responses of the CW-visitors in Chapters 4 and 5, as it provides a possible answer to what might happen if the CW-visitors stayed for longer and 'involved' themselves to a larger extent than what they do. Crossley shows that before volunteering

> encountering destitution is narrated as an unpleasant, yet necessary experience that one must go through in order to trigger emotions such as sadness and guilt, which in turn facilitate the positive change in the self.
> (Crossley 2012b, 94).

Interviews conducted later, while the volunteers are working with the beneficiaries, show that the volunteers are confused by the lack of unhappiness they encounter, leading some volunteers to believe that the locals hide the sadness that must be there as a consequence of their poverty, while others think they are experiencing genuine happiness in the locals despite their poverty, which leads them to question their assumptions of what constitutes poverty and what it means to the people affected by it. They tend to feel embarrassment when they can't hold back their tears on behalf of people who do not feel the tragedy the volunteers cannot help but project onto their lives. At home again, some volunteers remain unaffected and continue their lives as before, while others experience a 'reverse culture shock', where the initial shock of encountering poverty, conceptualised as a 'realisation', is followed by a period at home, where solidarity with poor people abroad is hard to act out in a 'Western' culture of affluence, as consumerism persists as a natural part of life 'at home'. An internal transformation has taken place, which is impossible to convert into external behaviour, resulting in frustration.

Crossley's study is important to the study of slum tourism because it suggests the existence of a series of 'scripts' of how representatives of the global North on the verge of adulthood is supposed to emotionally involve and subsequently detach themselves from an encounter with representatives from the global south as a part of the *bildung* a modern 'Grand Tour' should hold (Desforges 1998). It suggests that this 'script' of involvement contains the anticipation of what looks like a Benjaminian 'shock of the real' (Dovey and King 2012), and when this shock is not shared by the people whose everyday elicits it, the volunteers might become confused. Some might disavow the feeling of shock as inauthentic because of this, though others might feel the necessity to feel the shock *for* the people who have apparently become desensitised to the poverty that elicits it for the volunteer. Either way, Crossley concludes that moral maturity is seen as something you arrive at by virtue of having felt this shock at all. The studies conducted in this book shows that the CW-visitors generally do not expect to experience such a shock, with notable exceptions. In Chapter 3, Nick describes the transformative effect it had on him when he visited the CW for the first time in 2006, while in Chapter 5, Agneta seems to yearn for it, as she expects it to influence her and her children, whom she has brought for that express purpose.

Talking to SBT's volunteer and CW-coordinator, Poonam, it seemed that she was less enchanted both the added emotional 'involvement' of volunteers as opposed to CW-visitors, and she spoke of the initial shock of the volunteers as 'teething problems' – a bodily metaphor that recognised the very real pain experienced by them but also associated it with the inevitable liminal phase of passing into a state of maturity that the volunteers implicitly didn't possess yet. Poonam would encounter the effects of these 'teething problems' at weekly meetings with the volunteers, which she jokingly called 'cribbing sessions', and at several occasions she tried to make me, instead of her, attend them because she disliked them so intensely. At the meetings, the volunteers would typically complain about all the things they didn't understand or like about India, Delhi and the organisation. They were disorientated and frustrated that there weren't a well-defined set of assignments they could work on, that there was no one to assist them in the initial interactions with the children and had a sense that they had been cheated by the VSA that had mediated the contact to SBT. Apparently, the VSAs didn't make it sufficiently clear that of the $600 that volunteers automatically donated to the program as a fee for volunteering, only about $100 trickled down to SBT.

Poonam's reaction was that she, as a part of SBT's limited staff, should ideally assist SBT-children rather than volunteers and that they had to find their own way and make up their own minds about how to relate to the effects of poverty, one of which were limited resources and lack of planning. Furthermore, she pointed out that the reason the volunteers all used VSA's instead of volunteering directly via the SBT home page was that they wanted a large organisation from the global North, such as the VSA, to vouch for their volunteering experience – especially in case they needed to get formalised credits for it, in relation to their education.

She had grown tired of witnessing how empathy with India's poor turned into indignation with India's middle and upper classes in bouts of self-righteousness that all too seldom implicated the volunteers themselves. She refused to act as a facilitator of this initial state of their *bildung*, partly because what shocked them was her everyday life – understood not as the experience of poverty (as Poonam was middle class) but, rather, the chasm between the classes that is an integral feature of Indian society. Her reluctance also stemmed from her suspicion that the script they were acting out ended with them back home at the end of this *bildung* journey, older and wiser. But just like Crossley indicates, this 'wisdom' would most often consist of their realising that their emotional landscape of identification could not in the end include the bodies of the subjects they were crying for when Poonam encountered them. Their maturity was thereby supposed to stem from the emotional upheaval of encountering injustice and then realising it as a fact of the world that they could not fundamentally change.

Poonam's strategy was therefore to leave them alone for two weeks until the shock and the inevitable stomach upset had subsided, after which time she would put them to work until they went home. Poonam thus understood the added 'involvement' of volunteers, versus the decreased involvement on the part of CW-visitors, as equalling a script of emotional development the volunteers had to go through and holding their hand while they did so added up to her spending more resources, and she therefore much preferred the CW-visitors, who took the tour, donated and left her to do what she saw as her actual job: caring for the children of the shelter homes or generating funds to pay others to do so.

Interestingly, she seemed to encounter the same misconception that emotional engagement was more valuable than money, when communicating with some Indian donors. One day, I listened to one end of a conversation between Poonam and a representative from Barclays Bank, whose Delhi administration had decided that each of their Delhi-employees should work as a volunteer for SBT for one hour each, in order to broaden their minds, strengthen their ties with the local community and so on. Poonam explained to the representative that though Barclays Bank would pay the salary of their 200 employees for that hour, having them circulate through the non-governmental organisation's (NGO's) shelter homes to interact with the children would merely add to the general confusion, and that CW-visitors were in fact already paying good money for a similar experience in the shape of the CW-fee.

During my fieldwork, I witnessed how the volunteers responded to Poonam's attitude to their emotions and documented the trajectory of a young, German, female called 'Beate', who stayed at the volunteer flat for two months. I chatted with her most nights, wrote down what I observed and, finally, conducted a formal interview with her over Skype right after she had returned home. About a week after her arrival at SBT, she was put in charge of a boy called Panchhi, who was deaf and hadn't learned any kind of standardised sign language. He lived along with 150 other boys in the DRMC home, which was spacious but understaffed, and his linguistic abilities consisted of a series of rudimentary signs he seemed to have established as a means of communication with the transient figures that had populated his life until then. He had arrived in Delhi by train like most other children at the DMRC home, but his lack of communication skills meant that no one could be sure from where, and a series of Mowgli-like mythologies emerged about his past among the other children. Because of his slightly bulging eyes and a predilection for climbing trees and staying on a particular branch for hours he was given his nickname, Panchhi (Hindi: *bird*).

Beate felt it as quite a responsibility to be put in charge of this boy, which she found needed the care a parent would normally give. He sought physical contact to an excessive extent, and while this was common among SBT-children, it seemed exacerbated in Panchhi's case. This could also be because most of his communication consisted of him directing other people by physically touching other people, and so mere communication was thereby sometimes interpreted as affection.

Getting to know Panchhi took time. It's the basic condition of SBT-children who are taught by volunteers that they must keep track of their own education and metaphorically mark the place in the book where one volunteer has gotten to when he (or, more frequently, she) goes home so that the next one might pick up from the same place. This provides a patchy education for many, while for Panchhi it meant that he linguistically had to start anew with every volunteer, establishing signs for everything from toilet breaks to his name with each new teacher. As Beate arrived, another volunteer Panchhi had become attached to was leaving, and Beate's experience of caring for Panchhi was thus strung out between his need for care and understanding – in the most basic of senses – and the inevitable end that had been foreshadowed on the day she arrived. This lent a sense of urgency to her work that some of the other volunteers didn't have, and though she was also angry at her VSA and India generally like the other 'teething' volunteers, she felt important to Panchhi at least. Based on this feeling, she confided in me that she distinguished herself from the tourists who went on the City Walk and the short-term volunteers who weren't 'serious', especially since they broadcasted their actions on social media, which to her signified a lack of immersion in the field that she prided herself in.

About a month and a half into her stay, her sense of the long-term effects of her work was undermined somewhat. The Mowgli mythology surrounding Panchhi implied that he was a displaced child, separated from a context that had taught him his sign language, but in a country with 1.1 billion inhabitants, finding someone who knew him from before would be too much of a miracle to hope for. The miracle occurred, however. A woman visiting the DMRC home from a Christian organisation in the state of Karnataka 2,000 km away recognised him as a boy she had met before, and a touching scene ensued with the two hugging. An anticipation arose on Beate's part that Panchhi would finally escape the hell of perpetual misunderstanding as he would be reunited with parents or guardians, who would surely be waiting in anticipation for his return. As time went by, it became clear that no one within SBT had the resources to go to Karnataka to find the shelter home that Panchhi came from, especially since the woman implied that he might have run away from it by his own volition.

To Beate, this meant that her engagement with Panchhi was reduced to yet another temporary engagement that he wouldn't be able to tell anyone about, and this lowered her sense of self-importance. In the Skype interview I conducted immediately after she returned, she seemed to have resigned herself to thinking that she, at best, had provided some entertainment for the children and simply said, 'I could keep them happy for a while'. I asked if that was all, and she replied that what the children really needed were parents, and '[E]ven if I adopted one of the boys, I couldn't adopt all 150'. In the interview she is marked by *reverse culture* shock, relating how she spent half an hour the first morning home in her parents' house staring at her cupboard, overwhelmed by the amount of clothes she had at her disposal, after having spent two months with only three T-shirts that all looked the same to choose from. But as I asked why, it turned out that it wasn't meant as the comment on consumerism as in Crossley's case but, rather, a stunned longing for the simple immediacy of having the mission of making a few children happy, compared to the business of embarking on the life she had planned to live, symbolised by the enrolment in university which was to start shortly.

It became a metaphor for a 'real life', which somehow turned the volunteer experience into an adventure, meant to morally educate her about a reality that was no longer hers, even as it had been impressed on her that not feeling it as her own, while living it, was somehow frivolous. She had followed the script of involvement without any of the ironic distance identified by Chouliaraki and without capitalising on it by promoting her experience on social media, and now that same script dictated that she detach herself and shed the illusion that her added 'involvement' was for the benefit of anyone but herself.

My position within SBT

Was I different from Beate? In 2013 I was a white, 36-year-old male PhD student, who had spent all in all four years in India at the time and could speak some Hindi and Bengali, and while this set me apart from the other volunteers in terms of why I was there and what experience I could utilise when interacting with the SBT-children, I did share a set of conditions with my flatmates. My engagement was also temporary, and I also aimed at using my experiences within SBT to gain symbolic capital in the shape of a degree within the same, global meritocratic system of education as the other volunteers, and while everyone I interacted with was told I was writing a dissertation on SBT, most SBT-children began their interaction with me as if I was just another volunteer.

The DMRC shelter home is a large, sparsely furnished concrete building smelling of disinfectants, where a French organisation called the Sound School had left two giant aluminium trunks full of expensive musical instruments when funding ran out around 2011 and the teachers left. As I started teaching the handful of children who had maintained some of their skills on the instruments, it was hard not to see the abandoned instruments as a metaphor for the cycle of involvement and abandonment constituted by the succession of development projects and temporary engagement that swept over the SBT-children at regular intervals. Situated as a volunteer, I had to come to terms with the fact that I was now a part of that process and to ponder the ethical and methodological question of what positions I should attempt to inhabit in the forthcoming process of trust building, which ideally should lead to me gaining insight into the lives of my interlocutors.

Jane Reas (2015) writes about the harmful aspects of the practice of letting voluntourists temporarily engage with orphans without prior training in child care. Based on an extensive fieldwork in a Cambodian orphanage that largely ran on donations from voluntourists, Reas goes so far as to argue that what constitutes 'child abuse' should perhaps be expanded to include not only sexual and violent abuse but also the commodification of children's bodies taking place when orphanages in the global South feel they need to sell opportunities to care for their charges, without demanding that the voluntourists stay for any length of time or possess any of the skills that would be required in a similar setting in the global North. She points out that the children's clinginess towards strangers, which was seen as 'adorable' and welcoming by voluntourists in her case, was, in fact, a symptom of Reactive Attachment Disorder resulting from revolving door involvement and short-term care of voluntourists. She therefore concludes that the welcoming bodies and warm smiles of the children were what Scheper-Hughes (2002) would call 'commodities of last resort', sold by the children because they had little else to sell, with the orphanages and the marketing departments of travel agents as eager middlemen.

Chapter 4 of this book showed that CW-visitors like Jørgen were nervous that they were in fact buying such a commodity of last resort, and so I started observing the children at the DMRC home to see how they interacted with the myself and other adults. The newly arrived children were usually younger than the age of ten, and as I taught music to the older children, these newcomers tottered around in what I was told by the staff was sometimes merely a state of culture shock but often also included severe undernourishment, withdrawal symptoms of drug habits or the effects of mental illnesses. They would look for any kind of stable attachment among the overworked full-time employees, and when they encountered volunteers looking for a chance to make a difference to a child in need, like Beate did with Panchhi, some of them undoubtedly thought they had found such an attachment. It usually ended like Beate and Panchhi's relationship with the volunteers going home never to return, but as my interactions with Beate indicate, the more involved these volunteers became in the children they took care of, the less they treated them like a frivolous commodity they could purchase and discard after use.

The slightly older children who had settled in, been given new clothes and enjoyed the stability of regular meals, a bed of their own and the mild boredom of school would still bestow an overenthusiastic greeting on visitors, similar to the one described by Reas, but as they entered their teens a different pattern of behaviour would emerge, which resembled Scheper-Hughes's broader conception of 'reciprocity' in patron–client relationships (1993, 98–127) more than the exploitation suggested by Reas's use of Scheper-Hughes's later text (Scheper-Hughes 2002). To explain this, former volunteer- and CW-coordinator Jessie Hodges pointed out to me that most SBT-teens would either start avoiding volunteers or begin to use them for specific means while limiting their attachment to them. Among the latter group of teens, the ones who flourished were those who were disciplined enough to take control of their own learning to such an extent that they used the volunteers' knowledge to clear up specific challenges that they themselves identified. Among my most successful 'students' were for example the CW-guides who demanded that I take notes during their CWs and corrected their mother-tongue-influenced pronunciation or the music student who brought his own sheets of music to every class and told me what to teach him well in advance so that I would have time to prepare. Other students were less disciplined; some were heavily traumatised by their past experiences, and they clearly did not flourish in an environment of perennial temps.

So, while I agree with Reas that the short-term attachment of orphans to adults creates problems – whether they are absent parents, temporary staff or volunteers – and that a type of commodification coexists with other types of relationships within voluntourism, I am not convinced that it is merely the bodies of shelter home children that are sometimes commodified. To journalists and researchers, such as myself, performances of the children's personal stories are valuable as they can be transformed into 'stories' and 'data'. SBT-staff, performances documenting the SBT-children's progress provide SBT with a *raison d'être*. And as we learn from the interviews with the CW-visitors, all adults interacting with SBT-children find it important to inhabit an ethically legitimate position, as no one wants to be the one who inadvertently damages the children they are trying to help. What Jessie pointed to was that as the SBT-children grow older, they grow increasingly adept at commodifying these positions, performances and interactions, and to use them as objects of trade in affective economies, while simultaneously protecting themselves from the emotional difficulty of temporary involvement, by decreasing the attachment in the exchange. The narrative and performative techniques employed to reduce emotional attachment when narrating the personal story on the CW analysed in Chapter 6 of this book were thus part of a larger professionalisation of the relationship between SBT-children and temporary adults circulating through their lives.

During my fieldwork, I was wary that I was purchasing a 'commodity of last resort' by extracting data from children who could not refuse to interact with me, but I was, after all, more comfortable positioning myself within the affective economy established between SBT-children and the volunteers who were primarily there to teach them, rather than the affective economy established between volunteers posing as stable relations, who traded care for an attachment to younger children that they couldn't sustain. As my target group of 15- to 21-year-olds would not have accepted that I adopt that position in relation to them, I was cut off from this option in any case. But this also meant that I methodologically traded in the position of the impartial ethnographer, who gradually comes to be viewed as a member of the group he is trying to study by simply 'hanging out' (described in e.g. Hammersley and Atkinson 1995), for the position of a clearly demarcated outsider, whom the SBT-children could use for educational purposes, until such time that they felt comfortable confiding in me, which, in fact, happened after a few months.

Drawing on Venkatesh's ethnographic study of Chicago gangs (2009), one might think that offering to represent them in writing would have seemed alluring to SBT-ambassadors, but within SBT it was recognised that the attachment of reporters and academics looking for stories and research subjects is even more short-lived than that of volunteers and development programs, and the adolescents who were vying for positions as SBT-ambassadors seemed to have a clear sense that while volunteers were there to help, albeit for a short while, reporters and researchers were merely there to get their stories and data. This was illustrated by an interaction taking place in the volunteer flat shortly after Beate arrived. Iftekar sat with her and showed her a homepage dedicated to himself, where he had collected all the TV segments that showed him performing as a guide in news stories. Most of these were made in the wake of the movie *Slumdog Millionaire* (Swarup 2009) and came from Indian, British, American and Australian news agencies, which mostly adopted the movie's rags-to-riches perspective in their news segments about the CW-guides who were former street children. As the segments were played to Beate, they were accompanied by a steady stream of complaints from the retired guide next to her about how the news agencies had 'taken our story and left' (field notes). There was clearly an innocently flirtatious element to Iftekar showing off his previous interactions with the world press in the guise of him telling the female volunteer how misrepresented SBT was, but his dissatisfaction pointed to a challenge common to most SBT-ambassadors.

Iftekar came from a small village in the state of Uttar Pradesh to the east of Delhi but ran away when he was seven and started living with SBT in 1996. While there, he had not only managed to finish high school (12th standard) but had also had experiences that wouldn't have been granted him, had he stayed at home. There were symbolic events, like meeting then prime minister of Britain Tony Blair and his wife, who visited SBT in 2005; guiding ambassadors and delegates from countries in the global North on the CW; and acting in theatre performances attended by the cultural elite of Delhi. But joining the CW-program also afforded him daily opportunities to practise his English with visitors and volunteers and to pass a distance education program, and this culminated in him being chosen for an exchange program where he spent a year at an American university doing a light business course and getting a general introduction to the United States in 2012.

As I first encountered him, however, he had just returned from the United States, lived penniless with friends from SBT and tried to find employment without any other formal education than high school, while trying to support his brother's handbag manufacturing business. He remained fiercely loyal to SBT because of their sustained efforts to provide children like himself with a safe base, but the job placement they could offer him at Pizza Hut as a sales clerk was not what he had in mind for the future, and the phase was over where he could perform the role of the former street child as an SBT-ambassador.

As the months wore on and I followed his struggle towards finding formalised employment with informal qualifications, he one day, in passing, compared his performances at SBT to a small girl in a talent show, singing her heart out on the TV in the volunteer flat's common room. Her ability as a performer was only interesting to the public insofar as her disadvantaged position as a child with little training made her performance all the more impressive, and just like the real stars of talent shows are the judges, who stay on from season to season and facilitate the temporary rise of these 'stars of ordinariness', Iftekar had begun to ask questions of the facilitators of the performances, where he had been invited to perform as an ambassador for SBT. How interesting would his performances be to people whose agenda was not to be seen to help a former street child?

Seen within the theoretical framework of this book, Iftekar's earlier question might be posed this way: Could the cultural, social or symbolic capital acquired at the performances where he was asked to be an SBT-ambassador be converted into other forms of capital such as money? Was this conversion contingent on him perennially performing as a former street child? How, then, might the relation between different forms of capital be theorised as circulating within affective economies, and might this circulation be observed and described? Second, comparing Beate's and Iftekar's cases, how might their different positions within the affective economies of voluntourism be theorised?

Precariousness has been conceptualised as a set of material conditions that delineates a new vulnerable class called 'the precariat' (Standing 2011) which in an English context have been studied in large surveys based on socio-economic indicators (Savage et al. 2013). Job insecurity, coupled with extensive debt caused by student loans and the real-estate bubble, means that the financial crisis of 2007–8 has given rise to a set of conditions that in many cases, match Standing's definition but also associates the term very closely with the reconfiguring of class in the global North caused by the gradual dismantling of the welfare state. Focusing instead on the discursive construction of reality, Butler (2006) insists that precariousness in its fundamental sense pertains to the uncertainty of whether a subject's life is recognised as liveable and his or her death grieveable. Drawing on Levinas, Butler argues that suffering must be recognised by others as both visible and relevant for it to register as problematic and cause for action, and she illustrates this by pointing to post-9/11 America, where 'grieveability' was distributed in such a way that it was not extended to for example presumed terrorists. The distribution of grieveability to some others, but not other others, reveals or conceals their suffering, and it thereby also determines whether any action should be taken to alleviate this suffering. Precariousness thus not only is experienced but also granted visibility and legitimacy.[2]

In performances staged for donors, CW-guides, and SBT-ambassadors generally, align themselves with current street children in subject positions that have been conceptualised as 'liminal subalterns' in Chapter 2, but which in this perspective might also be called 'honorary subalterns'. They are invited to speak for the urban subalterns trapped within the informal sectors and spaces of Delhi, who have little means of self-representation within a discourse intelligible to the elite. The Indian state was never a welfare state, and so to call these urban subalterns an 'Indian precariat' might give the wrong connotations, but they do nonetheless inhabit a highly precarious position. As the SBT-ambassadors work towards illustrating this, they have to counter the possible invisibility of their suffering by performing their lives as liveable and their deaths or suffering as grieveable. Simultaneously, they must also represent the precariousness experienced by the urban subalterns they speak for as a condition that might, at least temporarily, be alleviated by the donations they are likely to receive from the donors they are performing for. This performance thereby enters into an affective economy, where affects connected to liveability, grieveability and precariousness are circulated in relation to money, as has been previously illustrated.

But SBT-ambassadors, such as CW-guides, don't just pocket the money that circulates towards them. On the CW and in SBT's aid performances generally, economic capital flows from donors to current or former street children via SBT, who uses it to provide facilities that are equally available to all their charges, such as shelter, food, education and so on. SBT-ambassadors thereby extract no economic capital from these exchanges, apart from that which is also made available to their peers. But do they accumulate social capital? To Bourdieu, social capital is

> the aggregate of the actual or potential resources which are linked to possession of a durable network of more or less institutionalized relationships of mutual acquaintance and recognition . . .
>
> (Bourdieu 2002 [1986], 282)

SBT-ambassadors are mainly given access to networks of privilege within performances, but these performances act as illustrations of the fact that an exchange of capital is possible to the children the SBT-ambassadors represent. And though the SBT-ambassadors' relationships with influential people such as the trustees are often characterised by both sides as containing 'mutual acquaintance and recognition' (Ibid.) it is a mutual recognition of the fact that they do not possess equal opportunities and encounter each other in performances where they are positioned as 'donating elite' versus 'honorary subalterns'.

Most of the capital SBT-ambassadors receive might thus be said to be *cultural*. It is, however, not mainly what Bourdieu would call *institutionalised cultural capital* or *symbolic capital* in the shape of degrees or formalised recognition from noteworthy institutions, and if the SBT-ambassadors do succeed in acquiring degrees from for example distance education programs, these never measure up to the degrees from international universities held by the elite they interact with during the performances, nor even usually the degrees from domestic universities attained by diligent members of the lower middle class. Nor is the capital they receive usually *objectified*, though they do sometimes have the use of objects associated with cultural capital, such as computers, expensive cameras, musical instruments or even fashion items, as when the fashion brand Benetton provided the CW-guides with uniforms. So, whereas Bourdieu would for example see it as an expression of lack of cultural capital to own a painting without mastering the discourse within which it is usually interpreted, the SBT-ambassadors are typically able to demonstrate mastery over objects of cultural capital that they do not, in fact, own, like playing instruments or taking photographs with equipment belonging to SBT. The form of cultural capital they possess most of is thus what Bourdieu calls *embodied* (Ibid.), and it exists in the shape of linguistic and performative abilities, as well as common points of reference with the elite. It is for example understood that while Iftekar can show Tony Blair around, he won't suddenly be granted a position within Blair's staff, and the question thus arises: How SBT-ambassadors might utilise the social and cultural capital they acquire?

To these forms of capital might also be added *erotic capital* (Hakim 2010) accumulated by SBT-ambassadors in interactions, such as the one between Iftekar and Beate, which was by no means unusual and sometimes led to temporary relationships of a somewhat secret nature. In terms of long-term relationships, the 2000s saw a wave of marriages between former SBT-children, and finding a partner within the organisation was described to me as important, since people who have not lived on the streets have a hard time understanding what it entails, and families who welcome a former street child as a future son- or daughter-in-law are few and far between. Performances, such as fundraising events, are thus important because of the opportunity they afford SBT-ambassadors of opposite gender to perform in front of each other – on- and offstage – and to thereby accumulate not only social capital, as Bourdieu defines it, but also erotic capital.

While the circulation and accumulation of all these forms of capital can be clearly identified in an Indian setting, the conversion between them and the access they grant their 'owners' are configured differently in India generally, as well

154 *The economy of resocialisation*

as within this specific case, as opposed to the French cases Bourdieu primarily analyses. One major difference is that India has extensive education and job reservation schemes for disenfranchised segments of the population based on caste, and there are thus plenty of chances for single individuals from these segments to gain entrance into positions that resemble those of the urban middle class, while often remaining separate from them, precisely because the urban middle class knows or suspects that this position has been gained by utilising reservations. The opposition to the idea of reservations is most often cast within a discourse of meritocracy, where the urban middle class attempts to maintain its position by gaining 'merit' within formalised systems of education. Chatterjee critiques this position however by pointing to the fact that it disregards how hard it is for disenfranchised segments to enter institutions that might grant them 'merit' and that the 'system

Figure 7.1 SBT-ambassador posing by a classic car at the Ojas Gallery after SBT's fundraising event

of public education catering to the rest of the population has been entirely dissociated from 'merit' (2013b, 269).

SBT-ambassadors, who like Iftekar has 'graduated' from SBT, find it difficult to experience upward social mobility that is not predicated on a performance of themselves as perpetually trying to escape the position of the 'former street child', while still associating themselves with it by positioning themselves as 'honorary subalterns. They thereby encounter a 'glass ceiling' that is similar to that experienced by other upwardly mobile representatives of Dalits in Indian society, who cannot escape the stigma of the quotas they are forced to utilise. This 'ceiling' is no less hard to penetrate than that described by Bourdieu as existing between the classes in France, but the absence of a systematised caste system and reservation schemes to counter it in France means that the dynamics that maintain this stratification are configured differently.

The researcher as performer?

In the evening of 19 March 2013, I found myself onstage along with a group of budding SBT-ambassadors at a fundraiser event in the beautiful garden of the Ojas Art Gallery, situated 200 m from the 12th-century landmark of Qutub Minar in wealthy South Delhi. The performers consisted of the music students from the DMRC shelter home, as well as a couple of latecomers, who had tagged along to the event without actually being able to play the songs we were performing, and the songs were consequently performed in a stumbling manner. After months of observing the CW, I was now both performing and facilitating the SBT-ambassadors' performance and thus had time to reflect. What did the performance consist of? Who was it performed for? How did it position me?

The audience seemed to have separated neatly into three groups occupying separate spaces on the lawn. Directly in front of the stage sat other SBT-ambassadors, who had been invited to the event, while the SBT-trustees and potential donors were seated farther away and waiters circulated between the two areas. From my vantage point on stage, I was reminded of the numerous End of Term Concerts I had attended as a music teacher, with parents seated in the back and schoolmates seated in the front, but in the affective economy existing between the performers and the different groups of listeners, economic, social and cultural forms of capital were being distributed according to a different set of parameters.

The potential donors had been picked from the economic and cultural elite that the SBT-trustees were a part of, and the event they had been invited to participate in featured a dance performance and a special screening of the movie *Salaam Bombay!*,[3] after which they were transferred to the immaculate gardens of the Ojas Gallery, which was owned by one the trustees. They passed his 1902 Ford automobile at the entrance – apparently only one of 50 other classic cars he was looking after – and were seated at circular tables facing a grand Neem tree and the impromptu stage. Once seated, a short documentary on the work of SBT was shown, after which the head of the board of trustees, Praveen Nair auctioned off a ride around Delhi in the classic Ford, which went for 5 lakhs ($8.00), and the potential donors were encouraged to buy signed copies of the newly published screenplay of *Salaam Bombay!* benefitting SBT. Then it was time for the musicians to perform.

Among the group of SBT ambassadors seated in front of the band were a group of male dancers who had performed earlier that evening. They were a couple of years older than the band and were performing as SBT-ambassadors regularly at national stages such as *Kala Bhavan* in Delhi's theatre district around *Mandi House*. With their trained bodies and confident demeanours they were much more interesting than the performing musicians to the female SBT-ambassadors seated at the same table, and as the performance progressed the girls' glassy smiles of politeness towards the stage were replaced by pained expressions eagerly awaiting the end to the music performance. One particular girl seemed to embody the humiliation of the musicians, as she had seated herself on the lap of one of the best-looking dancers and demonstratively put her hands over her ears, while eyeing the singer with a pained expression. From my position next to him, I could feel the humiliation, though I, of course, was positioned differently as the teacher, rather than the student, and did not enter into the circulation of erotic capital that was rapidly draining from the musicians on stage in relation to the SBT-girls.

By the fourth song one of the SBT-girls got up to sing a song in Hindi accompanied by the band. Her legs had been forcibly bent out of shape shortly after her birth, so as to prepare her for a life of begging, and the song was an allegory of this life, sung in the highly nasal register sometimes utilised by blind beggars in trains, now supposedly turned into an artistic expression performed on a stage. After a couple of stanzas, it seemed to register with Praveen Nair that the microphone of the band sounded much worse than the one used by the announcers, and with a wave of the hand, she stopped the band and exchanged the microphones. Instead of being connected to an old guitar amplifier, the female singer's voice now reached the ears of the donors through a crisp PA-system and earned her a modest applause, though the good microphone was handed back to the staff before the male singer entered the stage to sing the final song.

The performance ended, and we left the stage. The poor quality of the performance did not seem to add a discord to the evening as a fundraiser event, however, because seen in this perspective the SBT-ambassadors hadn't so much performed a set of classic rock songs as a display of their lives as liveable and their suffering as grieveable, which the intervention in the fourth song also highlighted. And while the guarded applause after the first few songs displayed the elites' recognition of the performers as visible, deserving receivers of aid, there seemed to be a general consensus that if they were expected to in fact listen to the performance as an aesthetic expression, then the performers needed to practise a bit more beforehand. Consequently, as the potential donors approached me after the concert – rather than the SBT-ambassadors – they did not comment on the quality of the music but, rather, thanked me 'for the effort' I had made as an unpaid teacher.

Even though the staff kept predictably silent during and after the performance, they played an important role before and after the performance, as a presence that reminded the SBT-ambassadors that this encounter between Delhi's elite and themselves as honorary subalterns was an anomaly. Had the venue staff worked in, say, Connaught Place, where businesspeople and beggars rub against each other at a regular basis, they would probably have been trained to be more vigilant in keeping out anyone but the elite. Still, the precarious position of the SBT-ambassadors meant that the venue staff were uneasy about what offers of hospitality they might extend to us. They were reluctant to let the official SBT bus enter the compound and to provide us with electricity for our instruments, and they downright refused to let us use the good microphone, which only Praveen Nair had the authority to lend us. They had correctly gauged that for the official economy of the fundraiser event to work, seen from the perspective of their employers, the SBT-ambassadors just had to play – they didn't have to play or sound well. And so, though the lack of applause from the waiters only conformed to conventions, their previous behaviour seemed to make their silence pregnant with a reminder of what would have happened if the band had been playing at a regular garden party as professionals, rather than as honorary subalterns.

This affected how I interpreted the lack of applause from the group of dancers and girls seated in front of the band. Like in an end-of-term concert, affect was circulated, accumulated and distributed in an economy that, in this case, seemed to rob the performers temporarily of their social and erotic capital in relation to their peers because of the poor quality of the performance. The fact that Praveen had intervened in the performance and given preference to a singer who better embodied the liveability and grieveability of all SBT-children, however, pointed to the fact that if the musicians wanted to be invited again and have decent microphones, they would have to inhabit this ideal too. It thus pointed to the donors as an important audience to please, not only in terms of the accumulation of social and cultural capital but also for the chance they might grant the SBT-ambassadors to accumulate erotic capital at future events in relation to other peers.

The waiters for their part succeeded in subtly upholding the social hierarchy that the event was supposed to be an exception from and thus served as a reminder for the SBT-ambassadors of the importance of pleasing donors and SBT-trustees, as they alone could provide them with opportunities to accumulate the social and cultural capital connected to being an SBT-ambassador. To me, however, the waiters' behaviour also became a symbol of the world Iftekar was trying to enter without performing as an SBT-ambassador, and brought home just how hard it would be for this new generation of SBT-ambassadors to convert the highly contingent forms of social capital and embodied cultural capital they had acquired by being SBT-ambassadors into other, more durable, forms such as economic or symbolic capital.

The attitude of the waiters, however, earned me the opportunity to show myself to be different from them. I was dressed in my all-purpose volunteer kit of jeans, sneakers and a hoodie, while the other SBT-ambassadors had made more of an effort to be presentable, but the venue staff still addressed me as the older, white male every time they had to give, or refuse, us permission to use their facilities, even though my Hindi was worse than that of both SBT-ambassadors and staff, and the process therefore was cumbersome and the position of power unwanted. When the guests started arriving they seated themselves, and the waiters immediately started distributing snacks. Thirty minutes later snacks had yet to materialise on the SBT-ambassador's table in front of the stage, but as I sat to talk to them, a waiter appeared behind me and tried to catch my eye for a mute approval of his serving me. He didn't know whether he would be fired for serving the honorary subalterns seated at the table or fired for not doing so, and he mutely pleaded for me to solve the conundrum by letting him serve me instead. He couldn't be blamed if these apparent interlopers consumed food that had been served for me. Turning from the waiter, I asked my companions whether they wanted something and thus signalled that I thought they were entitled to decide for themselves and that this was all the assurance I was going to give the waiter.

Treating the SBT-ambassadors the way I treated the donors felt good as it afforded me the opportunity to simultaneously perform an act of solidarity with the SBT-ambassadors and to 'call the bluff' of equality that the event was predicated on. But even as I rebuffed the waiter for instating the social hierarchy that the event was supposed to be an exception to, I was acutely aware that any anger I felt at this discrimination should perhaps be directed at the smiling, pleasant host of the Ojas Gallery who had trained his waiters to be unpleasant to interlopers so that he as his guests needn't perform that unpleasant task. Or perhaps at my own comfort that this space was blissfully free from the beggars haunting me as a white person staying in Delhi.

It also reflected back on the position I had put myself in as a researcher. In order to build the trust of 'gatekeepers' (Hammersley and Atkinson 1995, 63) to the community of SBT-ambassadors, who would give me access to spaces where they performed their identities as former street children, I had become a gatekeeper myself. I had become a volunteer, who trained the musicians to participate in performances that situated them as SBT-ambassadors, a figure of relative authority who helped them during the actual performances and who interceded on behalf of them when dealing with the staff, and this earned me not only their respect and tentative friendship but also the interest of other SBT-ambassadors, who saw possibilities of entering into a similar relationship with me. Furthermore, the work I put into teaching the students and the praise I received from the donors also made the SBT-staff more forthcoming towards me. I earned a spot on the guest list of subsequent fundraiser events that I would otherwise have paid a lot of money for, precisely because the admittance was a form of donation, and internal documents I had been trying to acquire and interviews I had been trying to schedule since I arrived two months previously were suddenly granted me.

I thus entered into an economy where former street children sold performances of liveability and grieveability in exchange for opportunities to circulate and possibly accumulate various forms of capital, which they couldn't 'spend', however, if they grew out of or distanced themselves from the subject position of 'honorary subaltern'. I thereby also became complicit in sustaining this economy, and my only solace was that the SBT-ambassadors I interacted with had long accepted this, and only seemed to act resentful towards volunteers, journalists and researchers who pretended things were otherwise.

Conclusion

The chapter showed that while volunteers value their added emotional involvement higher than the superficial involvement of slum tourists, representatives of aid organisations don't necessarily agree with this valuation. Facilitating volunteers' involvement takes effort that might otherwise be spent helping the intended benefactors, such as SBT-children, while slum tourists pay and leave the organisations to do their jobs relatively quickly, and this practice fits well within a logic that sees the interaction between them as an affective economy where affect is exchanged for money. Chapter 6 showed that CW-guides tell personal stories that don't invite transgressions of the boundary surrounding the space of comfortable affective negotiation, even if such a transgression is necessary to mount a structural critique of the causes of the poverty described. The two-hour format of the CW doesn't allow for the added engagement that would make such a critique relevant to voice, and so the guides prefer not to.

This chapter showed that SBT-staff and SBT-children of a certain age meet volunteers with a similar dispassionate focus on how they might help the organisation's benefactors in practical terms, rather than how the volunteers might 'involve' themselves, even though the volunteers sometimes dedicate months to working within SBT. The difference between CW-visitors and volunteers, however, seem to be that while the former with very few exceptions seem to know that their emotional involvement is unwelcome, there seem to exist an informal 'script' among volunteers that dictates both an initial involvement, which should ideally be shocking enough to leave an emotional mark, and a subsequent detachment. Apparently, this detachment is supposed to be born from a realisation that mounting a structural critique of the inequality experienced, which might translate into durable changes in the volunteers' lives at home, are naïve, impractical or both. The chapter showed that Poonam, as a representative of the SBT-staff that work towards alleviating the effects of this inequality, has little patience towards this naturalisation of inequality and thus prefers CW-visitors who pay what little they can but refrain from undergoing this emotional journey with herself and the SBT-children as reluctant participants.

CW-guides are part of a larger group of 'SBT-ambassadors' performing as honorary, liminal subalterns for SBT at functions, charity events and so on, and the chapter proceeded to focus on how this group extracts other forms of capital than money from performing at these events. Precariousness can be both a set of socio-economic conditions framed in sociological terms (Standing 2011) *and* a condition that might be granted visibility and thus has to be performed in order to be acted on by a public (Butler 2006). This theoretical double perspective is used to analyse a fundraiser event staged by SBT. The analysis shows that while performing the precarious subject position of an honorary subaltern, whose life is liveable and suffering/death is grieveable might seem to grant access to the accumulation of social, cultural and erotic capital, the conversion of these into more durable forms of capital that transcends the subject position of the honorary subaltern is difficult. It also shows that while it is possible to trace the circulation of affect, understood as affective 'im*press*ions', between the participants of such a performance, the circulation of economic, social and cultural capital can only be traced in such performances by referring to the socio-economic relations existing between the participants.

Since capital, like affect, only exists between bodies and 'ownership' of any type of capital has to be recognised in order for a conversion to take place, it is a matter of interpretation whether the transfer and accumulation of social and embodied cultural capital take place at all. The analysis shows that SBT-ambassadors believe that it does, though older SBT-ambassadors, 'like Iftekar', are sceptical of how it might be utilised. This observation becomes important in the theoretical framing of the relationship between volunteers and subalterns staying with NGOs, as these after a while seem to stop looking to the volunteers for stable relationships. It could be argued that it is thereby no longer the temporary engagement of volunteers itself that becomes problematic or necessarily the commodification and sale of the possibility for them to interact with subalterns. Rather, what is problematic is that the exchange of economic, social and cultural capital is constructed in such a way that even the honorary subalterns, who act as metonyms for a larger group of subalterns, never accumulate enough capital of any sort to truly escape this subject position.

Notes

1 A term whose genealogy reaches back to Said's (1978) use of the term *imaginative geographies* to describe the historically inaccurate division of the orient and the occident into spatially imagined opposites of each other.
2 We are in a sense back to Arendt's distinction between the politics of pity versus the politics of justice (see Chapter 5). Here the spectator might choose to inhabit the position of the good Samaritan who helps out of pity or a judge who judges whether the suffering is self-inflicted and thus whether it merits the help of others. Arendt, in her conceptualisation of these positions, in a sense, presupposes that the suffering is recognised *as* suffering, and this is where Butler's intervention is timely, as this is by no means always the case.
3 The movie that 'started' SBT – see Chapter 3.

Conclusion and further perspectives

As stated in the Introduction, this book has three aims: (1) to develop new approaches to slum tourism research, by using the perspectives of other academic disciplines to perform theoretical, methodological and analytical interventions; (2) to invite further studies of slum tourism as a practice within these neighbouring disciplines; and (3) to write an ethnography of a particular group of people working with slum tourism in Delhi, 2006–17.

The overview of chapters provided in the Introduction furthermore shows how each chapter performs this intervention, by drawing on particular disciplines. Each chapter ends with a conclusion that outlines its analytical results and reading these in succession traces one path through the material, that reflects how I conducted my study of SBT, and thus how I chronologically arrived at my analytical results. Instead of summarising this chronology once again, this overall conclusion instead traces two different paths through the book and seeks to reach a set of analytical results this chronology-based structure didn't make room for. It begins with the question of what might constitute a meaningful show/shield debate in the future if it isn't based on the visual metaphors of 'voyeurism' and 'poverty porn'. It then proceeds to outline the inner workings of the system of meritocracy subalterns might advance within by performing their own perpetually liminal subalternity and what limitations of transgression this system seems to hold.

The show/shield debate and (im)possible articulations of solidarity

The book begins by focusing on SBT as an organisation operating in relation to governmental bodies in Delhi in the 2000s and introduces these as a part of the show/shield debate surrounding slum tourism in general. This debate is continued by zooming in on the CW as a performance (Chapters 2 and 3), on the CW-visitors (Chapters 4 and 5), the CW-guides' (Chapter 6) and, finally, other groups within SBT such as volunteers, staff, trustees, SBT-'ambassadors' and myself as a researcher (Chapter 7). Theoretically, the book tries to move the show/shield debate away from accusations of 'voyeurism' and 'poverty porn', which tend to cast the commodification of poverty performances as a uniformly harmful and exploitative practice. Instead, it tries to analyse how these performances-turned-commodities enter economies, where they are exchanged for different kinds of capital that are circulated not only towards the 'poverty performers' but also towards other actors within the organisations facilitating these economies.

As Chapter 6 showed, entering into these economies might be emotionally harmful for poverty performers such as the CW-guides, but framing their poverty performances as commodities is a conscious strategy on their part in decreasing the emotional harm they might otherwise be exposed to. They prefer the professional distance of a business relationship, to an encounter defined as a 'human-to-human' meeting, as the latter dictates an emotional involvement, which would hurt without healing, because the relationship wouldn't be sustained beyond the two hours of the CW. Most slum tourists understand this, and the book concludes that friction between representatives from the global South and North on this account rather occurs in the encounter between SBT-children and volunteers, as the professionalised detachment of SBT-children conflicts with the script of involvement the volunteers feel obliged to follow.

As to the question of whether slum tourism is exploitative, the book tries to provide an answer, firstly by pointing to a series of similar practices that are generally not deemed exploitative above and beyond what is normally dictated by the global, capitalistic division of labour. If the majority of labourers working at leisure tourism destinations in the global South are placed within the same wage bracket as workers in slum tourism, then the possible exploitative nature of slum tourism must consist of something other than a skewed system of remuneration. The book then attempts to look for a structure in what type of poverty representation that has been derided as exploitative but concludes that almost any exchange might be accused of acting as poverty porn, if it includes signs relating to poverty circulating in affective economies resulting in someone's pleasure. Omitting the harsh realities of poverty in circumstances where they are really present might be viewed as soft poverty porn, while wallowing in descriptions of these harsh realities might be viewed as hardcore poverty porn.

Trying to make sense of this representational dilemma, the book relates it to the gradual undermining of representational strategies of humanitarian aid campaigns in the 20th and 21st centuries, and concludes that slum tourism might be understood not only as a tourist performance but also as an aid performance governed by a post-humanitarian logic. This logic is characterised by a disenchantment with yesteryears' grand ideological narratives of how global suffering might be alleviated by aid campaigns and because donating is decreasingly accompanied by a genuine, heartfelt belief that suffering is thereby alleviated in any long-term sense, the emotional investment in the suffering of distant others mediated by the aid performances has become increasingly temporary, ironic and contingent. This post-humanitarian logic is furthermore characterised by a conscious mixing of business and charity so that private gain and altruistic giving are no longer seen as mutually exclusive aims. In the case of the CW, this means that the CW-visitors are positioned as both consumers and donors, while the CW-guides are positioned as both workers and receivers of aid.

The difference between slum tourism performances, such as the CW, and post-humanitarian aid performances is that the potential receivers on the tours are not mediated to potential donors via screens, posters or textual descriptions. Rather, they enter into face-to-face encounters with each other, and this book tries to analyse what this physical presence does to the CW as a tourism/aid-performance? As mentioned earlier, the CW-guides react by downplaying their emotional involvement by emphasising their role as workers, rather than receivers of aid. Second, they invite the tourists to also downplay their emotional involvement in their past suffering by adopting an ironic stance towards it themselves. This seems to echo the ironic stance utilised in other post-humanitarian aid performances, where it acts as protection for prospective donors against representations of suffering that they cannot hope to fully alleviate, though they still feel morally obliged to try. The CW-guides furthermore try to validate the position of the CW-visitors by constructing personal stories for public consumption that clearly demarcate themselves and the other SBT-children as a separate group of potential receivers of aid, who can meaningfully be helped by relatively small donations via the framework of SBT, whereas the larger group of street children outside the organisation are discursively produced as beyond the reach of the CW-visitors. Hence, if the visitors believe the personal stories, they are liberated from the agony of trying to reach these children, or the frightening desensitisation that comes from repeatedly trying, but not being able to do so.

Returning to the question of what might constitute a meaningful show/shield debate, in a setting where guides try to pre-empt the moral discomfiture of slum tourists to protect themselves, the book first seeks to move away from the visual metaphor of the tourist gaze and instead towards a focus on tourist performances, enabling a discussion not of what tourists are shown on slum tours, but on how they *co-perform* them. It asks which kinds of interactions are encouraged or discouraged, why and by whom, and recognising that emotions play a significant part in determining this, the book proceeds to construct a framework where emotions might be registered as producing meaning in a social context. Consequently, the CW, and slum tours, in general, are theoretically framed as co-performed spaces of affective negotiation that hold areas of comfort and discomfort for the subjects performing them.

The preceding descriptions of the CW-guides' narrative and performative strategies can be understood as their attempts at shepherding the CW-visitors into areas that are comfortable to them, by situating themselves as workers rather than receivers of aid. But while they thereby protect themselves, they also foreclose the possibility that the visitors might meaningfully articulate declarations of solidarity that are not set within a post-humanitarian framework – a framework where aid performances are bought and sold for causes that only elicit moderate, temporary or ironic emotional responses. Some tourists attempt to transgress the boundaries delimiting this comfortable space. Sometimes because they are uneasy about the kind of post-humanitarian consumer-solidarity they are invited to declare in front of vulnerable others whom they believe deserve better, and at other times because they wish to declare their solidarity with a struggle against larger structures of inequality. The CW-guides, however, have no wish to engage in this struggle and consequently reject the visitors as politely as they can.

The analyses of Chapters 4 and 5 showed how uncomfortable this process is to the CW-visitors, both when they are on the CW and retrospectively during the interview, where the potential victims of their transgressive behaviour – be they guides or SBT-children – aren't even present. Jørgen tries to transgress the boundary retrospectively in the interview but ends up only touching it anxiously and shrinking from it again for fear of how the larger structures of commodification position him in relation to the guides and the SBT-children. Being an expatriate, Line has had time in India to recognise the hegemonic structures of meritocracy that the CW exists within, and she dislikes how positive emotions and 're-sensitisation' connected to charity are substituted for what she believes is real engagement. The CW's comfortable space of negotiation is thus actually uncomfortable for her, but she cannot transgress the boundaries delimiting it because rather than antagonising the elite she imagines is behind, she might instead antagonise the SBT-children and guides, who are trying to escape this repressive system by working within it. Agneta sees largely the same system as Line but feels complicit in having produced it, and anger thus turns to grief for her, but a grief that has a pedagogic aim. If she is complicit in upholding the system, then maybe she can change it by changing herself, and thus, she seems to amplify the emotional impact on her and her sons in order to bring about this change.

The results of this analysis suggest that future discussions of the ethical implications of slum tourism could profitably move away from the show/shield debate as it is conducted presently. They suggest that a focus on emotional interactions rather than 'sights' enables inquiries about the 'feeling rules' of slum tours as spaces of affective negotiation and thereby about what kinds of solidarities, struggles and emotional connections that are rendered impossible, impractical or even emotionally exploitative by these rules.

As the next section illustrates, the CW is a highly depoliticised slum tour, and researchers studying slum tours that are less so might well find that these invite other articulations of solidarity, which justify the commodification of poverty performances in different ways, enabling or foreclosing differently configured types of solidarity. Certainly, Frenzel (2016, chap. 9+10) makes this assertion, referencing examples from Mumbai and Rio de Janeiro in relation to logics of solidarity. This book, however, suggests that different perspectives emerge, if future studies of the possible articulations of solidarity on slum tours consider the affective side of the humanitarian logics these exist within, as well as the positions and actions these logics leave open to actors co-performing the tours.

This might also be of interest to researchers situated in the branch of development studies that concern itself with how appeals for aid to discursively constituted publics draw on and enable certain solidarities and etho-political visions. If, indeed, most slum tours promote themselves as both and tourist and aid performances, then slum tours are one of the rare cases were the suffering or vulnerable others normally portrayed in mediatised aid campaigns actually meet the potential donors they are supposed to appeal to. And while this book only shows how a small segment of visitors to a single slum tour relates to the appeals they were presented with, it does seem significant that the perceived necessity of an ironic stance, which is apparently a feature of both mediatised and face-to-face aid encounters, seems to be an emotional necessity not only for the potential donors and non-governmental organisations (NGOs) but also for the poverty performers themselves.

It seems to point to a general tendency that neither donors *nor* receivers want to sacrifice the feeling of comfort between them for an uncomfortable articulation of what concrete harm the inequality between them inflicts, even when it is that very inequality that is the theme of the encounter between them. A prospect that indicates that alternatives to a neo-liberal economy with sporadic acts of consumer solidarity might not be easy to applaud, let alone formulate, for the people who clearly lose in this economy, such as the Indian, urban subalterns living in *jhuggies* or precariously on the streets of wealthier inhabitants of megapolises. This is not to say that it is only the emotions circulating in the concrete interactions between subaltern and non-subalterns that institutes this inability to formulate alternate solidarities. The following section expands this point.

Subalternity, meritocracy and hegemony

A central figure throughout this book is that of the contemporary, urban subaltern. Chapter 1 theoretically framed inhabitants of slums and '*JJ colonies*' as urban subalterns, who have little chance of representing themselves within discourses intelligible or convincing to the elite, unless they represent themselves as pressure groups, constituencies or unions related to their subject positions as subalterns. Using Chatterjee's conceptual framework, one might say that they are trapped within the forcibly homogenised time-space of both *c*apital, understood as an agglomeration of wealth, and the *C*apital, understood as the heavily policed topos of Delhi.

In Chapters 3 and 6 the figure of the liminal, urban subaltern entered in the shape of the CW-guides. Chapter 3 showed that the CW-guides must inhabit this position if they are to represent both SBT and the current street children encountered on the CW, and Chapter 6 analysed the personal stories of the CW-guides, where this position is more explicitly articulated. Far from seeing the stories' highly codified, professionalised and indeed commodified nature as a mark of their 'inauthenticity', the chapter sees it as proof of their constant use as tools in a positioning game. A game that can be traced from the time when they first arrive at SBT, where they use their status as 'street children' to get access to services meant for this population group, to the time when they perform their 'personal stories' for donating CW-visitors. The chapter concludes that the CW-guides are situated ambivalently at the time of telling as former street children who are in the process of being resocialised, and this liminal position allows them to speak for current street children, while setting themselves apart from them. They are thereby positioned as liminal, urban subalterns, representing a group of street children whose fate they do not share *because* they have made a career representing them.

Using Chatterjee's analytical tools, the CW-guides can be said to occupy a position of subalternity in so far as they are forced to articulate a set of subject positions that are located within the population group '(former) street child', in order to get access to the services offered by SBT. But far from seeking political influence from this position, like for example the squatters in Akankasha Colony in Chapter 1, the guides seek to escape the position by becoming spokespersons for the subjects caught within subalternity. This escape is narrated as achievable for subjects who have acquired enough merit, and the CW-guides thereby validate the meritocratic system that claims to provide an escape route for them. Within this logic it would not make sense for the guides to imply that the donating foreign tourists and Indian elite are complicit in perpetuating the system that makes the deprivation the CW-guides face possible, not just because the guides might thereby antagonise potential donors but also because it would undermine the system of merit itself, and thereby their position as gifted achievers within it.

Chapter 7 explored to what extent an escape from the position of subalternity is actually achievable for SBT-children working within this meritocratic system. It develops this point with an ethnographic account of a fundraiser event arranged as an organised cultural encounter between potential donors and 'SBT-ambassadors'. The chapter outlined how different forms of capital is transferred between the participants at the event and how this is part of a larger economy, where SBT-ambassadors gain social capital and embodied forms of cultural capital in exchange for them performing the role of honorary subalterns. This validates not only the good work of SBT but also the larger meritocratic system that SBT acts within.

The whole point of the fundraiser event is to act as a temporary suspension of Delhi's social, and thereby spatial, division between the donating elite and this small group of former street children, who thereby become not only 'liminal subalterns' but also what I call 'honorary subalterns'. Subalterns who are supposed to be treated as equals for the duration of the event. But because both parties are so eager to show that this suspension is possible, and furthermore yearn to experience interactions, where the prevailing power relations between them are rendered temporarily imperceptible, the waiters at the event get confused. They are used to enforcing the division of space for an elite who pretends that it is not there, and because of this the distinction between ordinary urban subalterns and honorary subalterns remains opaque to them. They therefore accidentally enforce the social and spatial division that the event is supposed to be an exception to.

Both Chapters 1 and 7 thereby explored which bodies and objects are allowed to occupy elite spaces in the Indian metropolis and how this division of space is enforced, but in the latter chapter the division is not enforced by bulldozers but by the subtle social exclusion performed by the venue staff. Their presence comes to act as a reminder of what potentially awaits the SBT-ambassadors if they decide not to advance within the system of meritocracy set up by SBT.

Chatterjee shows how squatters can only gain political influence by positioning themselves as a population who should be given special entitlements but that they thereby also leave themselves open to being positioned as the pre-modern other of the nation state and a problem to be solved with tools of governance. He links this to how Dalits can get access to government quotas in order to for example attain degrees, but that these are not valued equally with those attained outside the quota system, and that the system of quotas that was set up in order to represent the actual heterogeneity of the nation perhaps thereby has ended up reinforcing the social divides it originally tried to bridge. The former street children analysed in this book seem to encounter dilemmas very similar to these about how to articulate their subject positions in different contexts, given the discourses that are made available to them. True to Gramsci's basic definition of hegemony (Gramsci 2011, 508–9), these discourses are framed, on one hand, by a system of coercion (bulldozers and waiters), and on the other hand a system of support they might advance within (performances within SBT), though this does not provide them with opportunities equal to the elite they interact with, and simultaneously erases the coercive side of this hegemonic system of meritocracy.

Future studies of poverty performances encountered within slum tourism or voluntourism to the global South might analyse issues of representation and incentive structures inspired by this conceptual framing. What is the relationship on other slum tours between poverty performers, such as guides or locals, and the people they claim to represent? Do we find similar authenticity constructions in Rio, Soweto, Mumbai or some of the many newer slum tour destinations? What kind of merit/capital do they gain by performing poverty, and might they lose it, if the step outside the system they have gained merit within, as in the case of the CW-guides? If so, how does it shape the performance? What forms of capital can the merit they have gained be exchanged for, and is their position as liminal subalterns truly liminal, in that it is possible to transcend it?

These are questions that might yield other analytical results if asked in settings where the slum tours conducted are more topos-driven than the CW. The CW departs from the format of other slum tours in that it doesn't take place in a slum, as the word is defined by for example the UN or the Delhi Development Authority, because the Akanksha slum was demolished in 2010 and the CW instead goes through the area of Pahar Ganj. The guides are relieved from having to worry about whether this will be demolished, though they have to work all the more at performing a perceived slum authenticity derived from their position as liminal subalterns. This is not the case on many other slum tours around the world, where the representational dilemma outlined in Chapter 1 about how visibility might lead to demolition is still pertinent. What representational strategies do guides working in slum areas that might be demolished because of their representation adopt? What merit or capital do the guides on these topos-driven tours attain, and at what cost?

The same questions might also be asked of practitioners in the field, such as NGOs facilitating encounters between subalterns and tourists, or agents working with such NGOs. One example is the many orphanages in the global South, who provide voluntourists the opportunity to temporarily care for their charges, as SBT did. If studies conducted in different settings yield similar results, then perhaps that should have consequences for how voluntourists are prepared by VSAs before they leave home and how NGOs use voluntourists in their daily work. This is, of course, assuming that these dynamics are alterable by pedagogic or organisational interventions and are not simply a product of a global inequality that has to be fundamentally dismantled before any solution might be found.

In terms of how neighbouring disciplines might benefit from this insight, it seems that the reluctance of upper- and middle-class India to see the practice of slum tourism as anything other than a reiteration of the colonial exotisation described in Chapter 3 is also shared by Indian scholars concerning themselves with the correlation between social and spatial exclusion in the capital. The literature cited in Chapters 1 and 2 describing the detrimental effects of modelling megapolises in the formerly colonised world to the ideal of the 'World Class City' of the formerly colonising world (e.g. Baviskar 2003, 2011; Baviskar and Ray 2011; Sundaram 2009; Sarda 2010) often allude to the existence of resistance to those same ideas in the global North, but they don't mention the fascination with informal urbanism among at least a segment of the 8 to 9 million foreign tourists visiting India each year.

While it is true that the itineraries of tourists travelling India exclude the vast majority of areas heavily affected by poverty by virtue of avoiding most of rural India, it is also true that an increasing number of development and conservation projects in urban areas are funded by visiting tourists. The disinterestedness of Indian scholars of social science to this fact disregards tourists as a potentially large economic force, that supports certain practices within the NGOs these scholars sometimes research and critique. It furthermore leaves the analysis of the cultural encounters taking place between Indians and foreigners in this setting to tourism researchers preoccupied with feasibility studies in a strictly economic sense. These are studies articulated within a neo-liberal episteme that many Indian scholars of social science are busy critiquing in other settings, and so an extension of their interests to include this perspective could be timely.

References

Adiga, Aravind. 2008. *The White Tiger*. London: Atlantic.
Ahmed, Sara. 2004a. *The Cultural Politics of Emotion*. London: Routledge.
Ali, Shri Sabir. 2010. "Urban Slums in Delhi." Directorate of Economics and Statistics. www.environmentportal.in/files/UrbanSlum_65thRound.pdf.
Anand, Mulk Raj. 1947. *Coolie*. London: Hutchinson International Authors.
Anderson, Benedict. 1991. *Imagined Communities: Reflections on the Origin and Spread of Nationalism*. Revised and extended. London: Verso. http://books.google.co.in/books?id=4mmoZFtCpuoC.
———. 1998. *The Spectre of Comparisons: Nationalism, Southeast Asia, and the World*. London: Verso.
Appadurai, Arjun, ed. 1988. *The Social Life of Things: Commodities in Cultural Perspective*. 1st Paperback Edition. Cambridge: Cambridge University Press.
Arendt, Hannah. 1990. *On Revolution*. Reprint. London: Penguin.
Bansal, Rashmi. 2012. *Poor Little Rich Slum*. 1st edition. Chennai: Westland.
Barenboim, Daniel, and Edward W. Said. 2002/1942. *Parallels and Paradoxes: Explorations in Music and Society*. New York, NY: Pantheon Books.
Batabyal, Somanth. 2010. "New Delhi's Times: Creating a Myth for a City." In *Finding Delhi: Loss and Renewal in the Megacity*, edited by Bharati Chaturvedi, 95–112. New Delhi: Penguin Books India.
Baviskar, Amita. 2003. "Between Violence and Desire: Space, Power, and Identity in the Making of Metropolitan Delhi." *International Social Science Journal* 55 (175): 89–98.
———. 2011. "Cows, Cars and Cycle-Rickshaws: Bourgeois Environmentalists and the Battle for Delhi's Streets." In *Elite and Everyman: The Cultural Politics of the Indian Middle Classes*, edited by Amita Baviskar and Raka Ray, 391–418. London: Routledge.
Baviskar, Amita, and Raka Ray. 2011. *Elite and Everyman: The Cultural Politics of the Indian Middle Classes*. London: Routledge.
Bell, Daniel. 1973. *Coming of Post-Industrial Society: A Venture in Social Forecasting*. New York, NY: Basic Books.
Benjamin, Walter, and Asja Lacis. 1942. "Naples." In *Reflections: Essays, Aphorisms, Autobiographical Writings*, edited by Walter Benjamin, 1892–1940. New York, NY: Schocken Books.
Benwell, B., and E. Stokoe. 2010. "Analyzing Identity in Interaction: Contrasting Discourse, Genealogical, Narrative and Conversation Analysis." In *The SAGE Handbook of Identities*, edited by Margaret Wetherell and Chandra Mohanty, 82–103. London: Sage.
Berg, Dag-Erik. 2017. "Race as a Political Frontier Against Caste: WCAR, Dalits and India's Foreign Policy." *Journal of International Relations and Development*, April: 1–24. doi:10.1057/s41268-017-0091-3.

References

Bhabha, Homi. 1984. "Of Mimicry and Man: The Ambivalence of Colonial Discourse." *October* 28: 125–33. doi:10.2307/778467.

———. 1990. "DissemiNation: Time, Narrative, and the Margins of the Modern Nation." In *Nation & Narration*, edited by Bhabha Bhabha, 291–322. London: Routledge.

———. 1994. *The Location of Culture*. New York, NY: Taylor & Francis.

———. 1998. "The White Stuff." Artforum International. www.questia.com/read/1G1-20757304/the-white-stuff.

Bhattacharyya, Deborah P. 1997. "Mediating India: An Analysis of a Guidebook." *Annals of Tourism Research* 24 (2): 371–89. doi:10.1016/S0160–7383(97)80007-2.

Birtchnell, Thomas. 2011. "Jugaad as Systemic Risk and Disruptive Innovation in India." *Contemporary South Asia* 19 (4): 357–72. doi:10.1080/09584935.2011.569702.

Blake, Stephen P. 2002. *Shahjahanabad: The Sovereign City in Mughal India 1639–1739*. Cambridge: Cambridge University Press.

Boltanski, Luc. 1999. *Distant Suffering: Morality, Media and Politics. Cambridge Cultural Social Studies*. Cambridge: Cambridge University Press.

Boo, Katherine. 2014. *Behind the Beautiful Forevers*. Random House Trade Paperback edition. New York, NY: Random House Trade Paperbacks.

Boorstin, Daniel Joseph. 1662. *The Image: A Guide to Pseudo-Events in America*. New York, NY: Vintage Books.

Bourdieu, Pierre. "The Forms of Capital." 2002 [1986]. In *Readings in Economic Sociology*, by Nicole Woolsey Biggart, 280–291. Blackwell Publishers Online.

Brooks, Peter. 1992. *Reading for the Plot: Design and Intention in Narrative*. 1. Harvard University Press edition. Cambridge, MA: Harvard University Press.

Bruner, Edward M. 2004. *Culture on Tour: Ethnographies of Travel*. 1st edition. Chicago: University of Chicago Press.

Burke, E. 1958. *A Philosophical Enquiry into the Origin of Our Ideas of the Sublime and Beautiful*. London: Routledge & Kegan Paul.

Butler, Judith. 1997. *Excitable Speech: A Politics of the Performative*. New York, NY: Routledge.

———. 2006. *Precarious Life: The Powers of Mourning and Violence*. London: Verso.

Certeau, Michel de. 1984. *The Practice of Everyday Life*. University of California Press.

Chakrabarti, Poulomi. 2008. "Inclusion or Exclusion Emerging Effects of Middle-Class Citizen Participation on Delhi's Urban Poor." *IDS Bulletin* 38 (6).

Chakrabarty, Dipesh. 2000. *Provincializing Europe: Postcolonial Thought and Historical Difference: Princeton Studies in Culture/Power/History*. Princeton, NJ: Princeton University Press.

Chatterjee, Partha. 1986. *Nationalist Thought and the Colonial World: A Derivative Discourse*. London: Zed Books.

———. 2004. *The Politics of the Governed: Reflections on Popular Politics in Most of the World*. Columbia University Press.

———. 2010a. *Empire and Nation: Essential Writings 1985–2005*. Ranikhet: Permanent Black.

———. 2010b. "We Have Heard This Before." In *Empire and Nation*, 267–75. India: Columbia University Press.

———. 2013a. *Lineages of Political Society: Studies in Postcolonial Democracy*. Columbia University Press.

———. 2013b. *Empire and Nation: Selected Essays*. Columbia University Press.

Chaturvedi, Bharati. 2010. *Finding Delhi: Loss and Renewal in the Megacity*. India: Penguin Books.

References

Chouliaraki, Lilie. 2010b. "Post-Humanitarianism: Humanitarian Communication Beyond a Politics of Pity." *International Journal of Cultural Studies* 13 (2): 107–26. doi:10.1177/1367877909356720.

———. 2012. *The Ironic Spectator*. Cambridge: Polity Press.

Clough, Patricia Ticineto. 1945–2007. *The Affective Turn: Theorizing the Social*. Durham: Duke University Press.

Conrad, Joseph. 1902. *Heart of Darkness*. Penguin Popular Classics. London: Penguin.

Crossley, Émilie. 2012a. "Poor But Happy: Volunteer Tourists' Encounters With Poverty." *Tourism Geographies* 14 (2): 235–53. doi:10.1080/14616688.2011.611165.

———. 2012b. "Affect and Transformation in Young Volunteer Tourists." In *Emotion in Motion: Tourism, Affect and Transformation*, edited by Dr. David Picard and Professor Mike. Robinson: Ashgate Publishing, Ltd.

Dalrymple, William. 1993. *City of Djinns: A Year in Delhi*. 1st edition. New Delhi: HarperCollins.

Davidson, Kelly. 2005. "Alternative India: Transgressive Spaces." In *Discourse, Communication, and Tourism*, 28–52. Tourism and Cultural Change, 5. Clevedon England: Channel View Publications.

Davis, Mike. 2006. *Planet of Slums*. London: Verso.

Delhi Department of Planning. 2001. "Economic Survey of Delhi 2001–2002." NewDelhi, India: Delhi Department of Planning.

Denzin, Norman K. 2008. "The Affective Turn: Theorizing the Social." *Contemporary Sociology*, Washington 37 (6): 604–5.

Desai, Kiran. 2006. *The Inheritance of Loss*. New York, NY: Atlantic Monthly Press.

Desforges, Luke. 1998. "Checking Out the Planet: Global Representations/Local Identities and Youth Travel." In *Cool Places: Geographies of Youth Cultures*, 175–92. London: Routledge.

Dickens, Charles. 1841. *Oliver Twist, Or, The Parish Boy's Progress*. Copyright edition. Leipzig: Tauchnitz.

Dirks, Nicholas B. 2001. *Castes of Mind: Colonialism and the Making of Modern India*. Princeton, NJ: Princeton University.

Dovey, Kim, and Ross King. 2012. "Informal Urbanism and the Taste for Slums." *Tourism Geographies* 14 (2): 275–93. doi:10.1080/14616688.2011.613944.

Dyson, Peter. 2012. "Slum Tourism: Representing and Interpreting 'Reality' in Dharavi, Mumbai." *Tourism Geographies* 14 (2):1–21. doi:10.1080/14616688.2011.609900.

The Economist. 2005. "Inside the Slums." *The Economist*, January. www.economist.com/node/3599622.

———. 2007. "Urban Poverty in India: A Flourishing Slum." *The Economist*, December 7. www.economist.com/node/10311293?story_id=10311293.

Edensor, Tim. 1998. *Tourists at the Taj: Performance and Meaning at a Symbolic Site*. Psychology Press.

———. 2001. "Performing Tourism, Staging Tourism (Re)Producing Tourist Space and Practice." *Tourist Studies* 1 (1): 59–81. doi:10.1177/146879760100100104.

Engels, Friedrich. 1872. *The Housing Question*. Moscow: Progress.

feministkilljoys. 2014. "Sweaty Concepts." *Feministkilljoys*, February 22. https://feministkilljoys.com/2014/02/22/sweaty-concepts/.

Foucault, Michel. 1972. *The Archaeology of Knowledge. (World of Man.)*. London: Tavistock Publications.

———. 1977. *Discipline & Punish: The Birth of the Prison*. Random House LLC.

———. 1978. *The History of Sexuality: An Introduction*. Pantheon Books.

———. 1980. *Power/Knowledge*, edited by Colin Gordon. Harvester Press.

Freire-Medeiros, Bianca. 2009. "The Favela and Its Touristic Transits." Geoforum, Themed Issue: The 'View From Nowhere'? *Spatial Politics and Cultural Significance of High-Resolution Satellite Imagery* 40 (4): 580–88. doi:10.1016/j.geoforum.2008.10.007.
———. 2013. *Touring Poverty*. London: Routledge.
Frenzel, Fabian. 2012. "Beyond 'Othering' The Political Roots of Slum-Tourism." In *Slum Tourism: Poverty, Power and Ethics*, edited by Ko Koens, Malte Steinbrink, and Fabian Frenzel, 49–66. London: Routledge.
———. 2016. *Slumming It: The Tourist Valorization of Urban Poverty*. 1st edition. London: Zed Books.
Frenzel, Fabian, and Ko Koens, eds. 2014. *Tourism and Geographies of Inequality: The New Global Slumming Phenomenon*. 1st edition. London: Routledge.
Frenzel, Fabian, Ko Koens, and Malte Steinbrink. 2012. *Slum Tourism: Poverty, Power and Ethics*. London: Routledge.
Frenzel, Fabian, Ko Koens, Malte Steinbrink, and Christian M. Rogerson. 2015. "Slum Tourism: State of the Art." *Tourism Review International* 18 (4): 237–52. doi:10.3727/1 54427215X14230549904017.
Gandhi, M. K. 1982. *An Autobiography: Or the Story of My Experiments With Truth*. Repr. Harmondsworth: Penguin.
Gentleman, Amelia. 2006. "Slum Tours: A Day Trip Too Far?" *The Guardian*, May 6, sec. Travel. www.theguardian.com/travel/2006/may/07/delhi.india.ethicalliving.
Ghertner, D. Asher. 2011. "Rule by Aesthetics: World-Class City Making in Delhi." In *Worlding Cities: Asian Experiments and the Art of Being Global*. London: Oxford: Wiley-Blackwell.
Gill, Kaveri. 2012. *Of Poverty and Plastic: Scavenging and Scrap Trading Entrepreneurs in India's Urban Informal Economy*. Oxford: Oxford University Press.
Goffman, Erving. 1959. *The Presentation of Self in Everyday Life*. A Doubleday Anchor Original. Garden City, NY: Doubleday.
———. 2008. *Behavior in Public Places: Notes on the Social Organization of Gatherings*. New York, NY: The Free Press.
Goldberg, David Theo. 1993. *Racist Culture: Philosophy and the Politics of Meaning*. Oxford: Wiley-Blackwell.
Graeber, David. 2001. *Toward an Anthropological Theory of Value: The False Coin of Our Own Dreams*. New York, NY: Palgrave Macmillan.
Gragnolati, Michele, Meera Shekar, Monica Das Gupta, Caryn Bredenkamp, and Yi-Kyoung Lee. 2005. "India's Undernourished Children: A Call for Reform and Action." https://openknowledge.worldbank.com/handle/10986/13644.
Gramsci, Antonio. 1971. *[Quaderni del carcere.] Selections From the Prison Notebooks of Antonio Gramsci*, edited and translated by Quintin Hoare and Geoffrey Nowell Smith. Translated by Quintin Hoare and Geoffrey Nowell SMITH. London: Lawrence & Wishart.
———. 2011. *Prison Notebooks*. Columbia University Press.
Gramsci, Antonio. 2005 [1926].*The Southern Question*. Translated by Pasquale Verdicchio. Canada: University of Toronto Press.
Guha, Ramachandra. 2008. *India After Gandhi: The History of the World's Largest Democracy*. Indian edition. India: Picador.
Gupta, Dipankar. 1997. *Rivalry and Brotherhood, Politics on the Life of Farmers in North India*. New Delhi: Oxford University Press.
———. 2000. *Mistaken Modernity: India Between Worlds*. India: HarperCollins Publishers.

References

Hakim, Catherine. 2010. "Erotic Capital." *European Sociological Review* 26 (5): 499–518. doi:10.1093/esr/jcq014.
Hall, Stuart. 2001. "The West and the Rest." In *Formations of Modernity*, edited by Stuart Hall and B. Gieben, 257–330. Milton Keynes: Open University Press and Wiley-Blackwell.
Hammersley, Martyn, and Paul Atkinson. 1995. *Ethnography: Principles in Practice*. 2nd edition. London: Routledge.
Hannah Arendt. 1958. *The Origins of Totalitarianism*. (Second Enlarged Edition [of "The Burden of Our Time"].). London; printed in USA: George Allen & Unwin.
Hannam, Kevin, and Anya Diekmann. 2011. *Tourism and India: A Critical Introduction*. London: Routledge.
Harvey, David. 2001. *Spaces of Capital: Towards a Critical Geography*. New York, NY: Routledge.
———. 2012. *Rebel Cities: From the Right to the City to the Urban Revolution*. London: Verso.
Hochschild, Arlie Russell. 1983. *The Managed Heart: Commercialization of Human Feeling*. 20th anniversary edition. Berkeley, CA: University of California Press.
Hodges, Jessie. 2010. "City Walk: Guide Training Manual." SBT, internal document.
———. 2011. "Development Smart- Acknowledging the Power That Street Children Bring to the Development Interface." MA Social Anthropology of Development, London, England: School of Oriental and African Studies, University of London.
Höijer, Birgitta. 2004. "The Discourse of Global Compassion: The Audience and Media Reporting of Human Suffering." *Media, Culture & Society* 26 (4): 513–31. doi:10.1177/0163443704044215.
Holst, Tore. 2012. "Parallel Lines and Lives: Edward Said as a Musician." In *Another Life/ Une Autre Vie*, edited by Vilain and Misrahi-Barak. Montpellier, France: Presses Universitaires de la Méditerranée PULM.
Holst, Tore, and Suranjita Mukherjee. 2009. *Turen Går Til Sydindien*. Politikens Turen Går Til. Kobenhavn: Politiken.
———. 2011. *Turen Går Til Nordindien*. Politikens Turen Går Til. Copenhagen: Politiken.
Hudson, Chris. 2010. "Delhi: Global Mobilities, Identity, and the Postmodern Consumption of Place." *Globalizations* 7 (3): 371–81. doi:10.1080/14747731003669750.
Hutnyk, John. 1996. *The Rumour of Calcutta: Tourism, Charity and the Poverty of Representation*. London: Zed Books.
Jain, P. 2016. "Rethinking the Scope of South-South Co-Operation in the Wake of Attacks on Africans in India." Africa at LSE. http://blogs.lse.ac.uk/africaatlse/2016/08/03/rethinking-the-scope-of-south-south-co-operation-in-the-wake-of-attacks-on-africans-in-india/
Johnson, Eleanor K. 2015. "The Business of Care: The Moral Labour of Care Workers." *Sociology of Health & Illness* 37 (1): 112–26. doi:10.1111/1467-9566.12184.
Kapoor, Dip. 2005. "NGO Partnerships and the Taming of the Grassroots in Rural India." *Development in Practice* 15 (2): 210–15. doi:10.1080/09614520500041864.
———. 2007. "Subaltern Social Movement Learning and the Decolonization of Space in India." *International Education* 37 (1): 10–41.
Kapoor, Ilan. 2004. "Hyper-Self-Reflexive Development? Spivak on Representing the Third World 'Other.'" *Third World Quarterly* 25 (4): 627–47.
Khair, Tabish. 2000. "The Fissured Surface of Language in Indian English Poetry." *PN Review* 26 (4): 7.

Koens, Ko. 2012. "Competition, Cooperation and Collaboration." In *Slum Tourism: Poverty, Power and Ethics*, edited by Fabian Frenzel, Malte Steinbrink, and Ko Koens, 32: 83. www.google.com/books?hl=da&lr=&id=vn4MOc570TYC&oi=fnd&pg=PA83&dq =competition+collaboration+ko+koens&ots=jBixwzZJiV&sig=FoGPlOeRXyzIF4MP2 JLt9B4wZBw.

Koven, Seth. 2004. *Slumming: Sexual and Social Politics in Victorian London*. Princeton: Princeton University Press.

Kvale, Steinar. 1996. *Interviews: An Introduction to Qualitative Research Interviewing*. Thousand Oaks, CA: Sage.

Leys, Ruth. 2011. "The Turn to Affect: A Critique." *Critical Inquiry* 37 (3): 434–72. doi:10.1086/659353.

Ma, Bob. 2010. "A Trip into the Controversy: A Study of Slum Tourism Travel Motivations." http://repository.upenn.edu/cgi/viewcontent.cgi?article=1011&context= uhf_2010.

Macaulay, Thomas Babington. 1935. *Speeches by Lord Macaulay, With His Minute on Indian Education. Selected With an Introduction and Notes by G. M. Young.* [World's Classics.]. London: Oxford University Press.

MacCannell, Dean. 1976. *The Tourist: A New Theory of the Leisure Class*. University of California Press.

Mayo, Katherine. 1933. *Mother India.* (4. impr.). Florin Books, No 19. London, Toronto: Jonathan Cape.

Meged, Jane Widtfeldt. 2010. *The Guided Tour: A Co-Produced Tourism Performance*. Department of Environmental, Social and Spatial Changes, Roskilde University.

Meschkank, Julia. 2011. "Investigations into Slum Tourism in Mumbai: Poverty Tourism and the Tensions Between Different Constructions of Reality." *GeoJournal* 76 (1): 47–62. doi:10.1007/s10708-010-9401-7.

———. 2012. "Negotiating Poverty: The Interplay Between Dharavi's Production and Consumption as a Tourist Destination." In *Slum Tourism: Poverty, Power and Ethics*, edited by Fabian Frenzel, Ko Koens, and Malte Steinbrink. London: Routledge.

Miller, Sam. 2010. *Delhi: Adventures in a Megacity*. 1st edition. New York, NY: St. Martin's Press.

Miller, William Ian. 1997. The Anatomy of Disgust. Cambridge, Mass: Harvard University Press.

Mohanty, Sachidananda. 2003. *Travel Writing and the Empire*. Katha.

Morrison, Toni. 1992. *Playing in the Dark, Whiteness and the Literary Imagination*. Cambridge, MA, London: Harvard University Press.

Muccino, Gabriele. 2006. *The Pursuit of Happyness*. Biography, Drama.

Mukherjee, Meenakshi. 2000. *The Perishable Empire: Essays on Indian Writing in English/Meenakshi Mukherjee*. New Delhi, Oxford: Oxford University Press.

Naipaul, V. S. 1967. *The Mimic Men*. London: Deutsch.

Nair, Mira. 1988. *Salaam Bombay!* Crime, Drama.

———. 2001. *Monsoon Wedding*. Comedy, Drama, Romance.

———. 2007. *The Namesake*. Drama.

———. 2013. *The Reluctant Fundamentalist*. Thriller.

Nyers, Peter. 2003. "Abject Cosmopolitanism: The Politics of Protection in the Anti-Deportation Movement." *Third World Quarterly* 24 (6): 1069–93.

Pandey, Gyanendra. 1989. "'The Colonial Construction of Communalism: British Writings on Banaras in the 19th Century." *Subaltern Studies* VI (VI).

———. 2006. *The Construction of Communalism in Colonial North India*. 2nd edition. Oxford: Oxford University Press.

Perkins, Harvey C., and David C. Thorns. 2001. "Gazing or Performing? Reflections on Urry's Tourist Gaze in the Context of Contemporary Experience in the Antipodes." *International Sociology* 16 (2): 185–204. doi:10.1177/0268580901016002004.

Picard, David, and Mike Robinson. 2012. *Emotion in Motion: Tourism, Affect and Transformation*. Ashgate Publishing, Ltd.

Potter, Jonathan, and Margaret Wetherell. 1987. *Discourse and Social Psychology: Beyond Attitudes and Behaviour*. Sage.

Pratt, Mary Louise. 1992. *Imperial Eyes: Travel Writing and Transculturation*. London: Routledge.

Puri, Jyoti. 2017. "Sexual States." Duke University Press, June 2. www.dukeupress.edu/sexual-states.

Radjou, Navi, Jaideep Prabhu, and Simone Ahuja. 2012. *Jugaad Innovation: Think Frugal, Be Flexible, Generate Breakthrough Growth*. 1st edition. Jossey-Bass.

Reas, P. Jane. 2015. "'So, Child Protection, I'll Make a Quick Point of It Now': Broadening the Notion of Child Abuse in Volunteering Vacations in Siem Reap, Cambodia." *Tourism Review International* 18 (4): 295–309. doi:10.3727/154427215X14230549904170.

Richey, Lisa Ann, and Stefano Ponte. 2011. *Brand Aid: Shopping Well to Save the World*. A Quadrant Book. Minneapolis: University of Minnesota Press.

Roberts, Gregory David. 2003. *Shantaram: A Novel*. Palgrave Macmillan.

Rogerson, Christian M. 2004. "Urban Tourism and Small Tourism Enterprise Development in Johannesburg: The Case of Township Tourism." *GeoJournal* 60 (3): 249–57.

Roy, Arundhati. 1997. *The God of Small Things*. London: Flamingo.

Said, Edward W. 1978. *Orientalism*. New York, NY: Routledge & Kegan Paul, Vintage Books.

———. 1993. *Culture and Imperialism*. New York, NY: Knopf.

———. 1994. *Orientalism*. Reprint. New York, NY: Vintage Books.

———. 2006. *On Late Style*. London: Bloomsbury.

Sainath, Palagummi. 1996. *Everybody Loves a Good Drought: Stories From India's Poorest Districts*. Penguin Books.

———. 2011. "India Together: Some States Fight the Trend, But Still. . . : P Sainath – December 8." http://indiatogether.org/suidata-op-ed.

Sarbin, Theodore R. 1986. *Narrative Psychology: The Storied Nature of Human Conduct*. New York, NY: Praeger.

Sarda, Shveta. 2010. *Trickster City: Writings From the Belly of the Metropolis*. New Delhi: Penguin.

Savage, Mike, Fiona Devine, Niall Cunningham, Mark Taylor, Yaojun Li, Johs Hjellbrekke, Brigitte Le Roux, Sam Friedman, and Andrew Miles. 2013. "A New Model of Social Class? Findings From the BBC's Great British Class Survey Experiment." *Sociology* 47 (2): 219–50. doi:10.1177/0038038513481128.

Scheper-Hughes, Nancy. 1993. *Death Without Weeping: The Violence of Everyday Life in Brazil*. Paperback print. Berkeley, CA: University of California Press.

———. 2002. "The Ends of the Body – Commodity Fetishism and the Global Traffic in Organs." *SAIS Review* 22 (1): 61–80.

Selinger, Evan, and Kevin Outterson. 2011. "The Ethics of Poverty Tourism." *Environmental Philosophy* 7 (2): 93–114.

Sengupta, Mitu. 2010. "A Million Dollar Exit from the Anarchic Slum-World: Slumdog Millionaire's Hollow Idioms of Social Justice." *Third World Quarterly* 31 (4): 599–616. doi:10.1080/01436591003701117.

Sharma, Kalpana. 2000. *Rediscovering Dharavi: Stories From Asia's Largest Slum*. India: Penguin Books.

Simpson, Kate. 2004. "'Doing Development': The Gap Year, Volunteer-Tourists and a Popular Practice of Development." *Journal of International Development* 16 (5): 681–92. doi:10.1002/jid.1120.

Sinha, Mrinalini, 1960–2006. *Specters of Mother India: The Global Restructuring of an Empire: Radical Perspectives*. Durham: Duke University Press.

Singh, Sarina, Mark Elliott, Abigail Hole, Kate James, Anirban Mahapatra, Daniel McCrohan, John Noble, and Kevin Raub. 2011. *Lonely Planet India*. Footscray, Victoria, Australia: Lonely Planet.

Smith, Adam. 1793. *The Theory of Moral Sentiments: Or, an Essay Towards an Analysis of the Principles by Which Men Naturally Judge Concerning the Conduct and Character, First of Their Neighbours, and Afterwards of Themselves: To Which Is Added, a Dissertation on the Origin of Languages: A New Ed*. 2 Vol. Basil: J. J. Tourneisen.

Spivak, Gayatri Chakravorty. 1987. *In Other Worlds: Essays in Cultural Politics*. New York, NY: Methuen.

———. 1988. "Can the Subaltern Speak?" In *Marxism and the Interpretation of Culture*. Champagne, Ill: University of Illinois Press.

———. 1993. *Outside in the Teaching Machine*. New York, NY: Routledge.

———. 2003. *Death of a Discipline: The Wellek Library Lectures in Critical Theory*. New York, NY: Columbia University Press.

Standing, Guy. 2011. *The Precariat: The New Dangerous Class*. London: Bloomsbury Academic.

Steinbrink, Malte. 2012. "'We Did the Slum!' – Urban Poverty Tourism in Historical Perspective." *Tourism Geographies* 14 (2): 213–34.

———. 2014. "Festifavelisation: Mega-Events, Slums and Strategic City-Staging – The Example of Rio de Janeiro." *DIE ERDE – Journal of the Geographical Society of Berlin* 144 (2): 129–45. doi:10.12854/erde.v144i2.59.

Steinkrüger, Jan-Erik. 2016. "Slums Als Thematisierung. Das Beispiel Shanty Town in Bloemfontein." *Zeitschrift Für Tourismuswissenschaft* 6 (2): 243–54. doi:10.1515/tw-2014-0210.

Sundaram, Ravi. 2009. *Pirate Modernity: Delhi's Media Urbanism*. New York, NY: Routledge.

Swarup, Vikas. 2009. *Slumdog Millionaire*. New edition. London: Black Swan.

Urry, John. 1990. *The Tourist Gaze: Leisure and Travel in Contemporary Societies*. London: Sage.

Urry, John, and Jonas Larsen. 2011. *The Tourist Gaze 3.0*. Sage.

Venkatesh, Sudhir. 2009. *Gang Leader for a Day: A Rogue Sociologist Crosses the Line*. London: Penguin.

Verma, Gita Dewan. 2002. *Slumming India: A Chronicle of Slums and Their Saviours*. New Delhi: Penguin Books.

Vodopivec, Barbara, and Rivke Jaffe. 2011. "Save the World in a Week: Volunteer Tourism, Development and Difference." *European Journal of Development Research* 23 (1): 111–28.

Wetherell, Margaret. 2007. "A Step Too Far: Discursive Psychology, Linguistic Ethnography and Questions of Identity." *Journal of Sociolinguistics* 11 (5): 661–81. doi:10.1111/j.1467-9841.2007.00345.x.

Wetherell, Margaret, and Jonathan Potter. 1992. *Mapping the Language of Racism: Discourse and the Legitimation of Exploitation*. New York, NY: Columbia University Press.

Wright, Carroll Davidson. 1894. *The Slums of Baltimore, Chicago, New York, and Philadelphia: Prepared in Compliance with a Joint Resolution of the Congress of the United States*, Approved July 20, 1892. U.S. Government Printing Office.

Appendix

Map 2.1 Map of the old CW-route. (a) Meeting Point pre 2010; (b) Akanksha Colony; (c) Community work at platform; (d) Platform where street children sleep; (e) GRP; (f) Pottery market; (g) Old haveli; (h) Aasra Shelter Home, SBT

180 *Appendix*

Map 2.2 Map of the new CW-route. (a) Meeting Point in 2013; (b) Small idols by tree and animal feeding; (c) 1st stop at car park; (d) Lots of trash; (e) 2nd stop: Recycle Shop; (f) former mosque; (g) 3rd stop: God Lane; (h) 4th stop: Main Bazaar; (i) Crossing the Chelmsford Road; (j) 5th stop: GRP Contact Point; (k) Crossing the Chelmsford Road; (l) Food stalls, (m) Pottery market, (n) Old haveli; (o) Aasra Shelter Home, SBT

Index

Adiga, Aravind 39
affect *vs.* emotions 87–8
affective economy 90; and capital (*see* Bourdieu, Pierre)
affective turn 87–8
Ahmed, Sara 4–5, 66–79, 87–95, 113, 124, 137–8
Akanksha Colony 30–5, 179
Anand, Mulk Raj 37
Anderson, Benedict 22–3, 64
anger 102–3
anxiety 95–9; *vs.* fear 95–6
Appadurai, Arjun 38
appeals 103–8, 143, 164
Arendt, Hannah 104–5, 112, 143
authenticity 197; *vs.* authentication 90; of guides 47–55, 78; of slums 67; in tourism 42–4

Bansal, Rashmi 39
Barenboim, Daniel 22
Batabyal, Somanth 39
Baviskar, Amita 31, 37–40, 68, 65, 168; bourgeois environmentalism 26–7
Bell, Daniel 91
Benjamin, Walter 4, 23, 66, 144
Bhabha, Homi 4, 21–3; whiteness 61–4
Bhattacharyya, Deborah 46
Birtchnell, Thomas 40
Blake, Stephen 46
Boltanski, Luc 104, 106, 113
Boo, Katherine 39
Boorstin, Daniel 42–3
bound serialities 22–3
Bourdieu, Pierre 152–5
bourgeois environmentalism *see* Baviskar, Amita
Brooks, Peter 125

Bruner, Edward 41–4; postcolonial tourism 61–2; tourist talk 87–8
Burke, E. 67
Butler, Judith 152, 159

caste *see* Dirks, Nicholas
Chakrabarti, Poulomi 26
Chakrabarty, Dipesh 36
Chatterjee, Partha 3, 19–27, 31, 37–42, 50, 55, 64–5, 68, 124, 154, 165–6
Chouliaraki, Lilie 5, 104–8, 112, 143, 147
City Walk description 48–54, 65–78, 93–6; critique 11, 33; feedback forms 83–4
Clough, Patricia 87
comfort 72; comfortable space 135–9, 163–5; tourists' discomfort 89, 96–9, 108–18
Conrad, Joseph 12

Dalrymple, William 39
Davidson, Kelly 46, 51, 72
Davis, Mike 15–17; discussion with Sundram 40
Denzin, Norman 87
Desai, Kiran 1
Desforges, Luke 144
Dickens, Charles 128, 131
Dirks, Nicholas 24, 64
disgust 70–1
distant suffering 104
Dovey & King 4, 17, 66–7, 71, 103
Dyson, Peter 2, 18, 44, 67

Economist, The 39–40
Edensor, Tim 41–2, 60

Index

emotional labour 91–2, 99, 109–11, 132–9
empathy 106, 113–14
Engels, Friedrich 15–18, 137

Feministkilljoys 69
Foucault, Michel: *vs.* de Certeau 56; postcolonial studies 21; tourist studies 41; whiteness 59, 62, 67–8
Freire-Medeiros, Bianca 2
Frenzel, Fabian 66; othering or 'same-ing' of slum inhabitants 18, 69; resistance to slum tourism 101–3, 137, 164; tourist valorization 44, 89–92

Gandhi, Mohandas 24, 63–4, 80
Gentleman, Amelia 11, 33
Ghertner, D. Asher 27, 68
Gill, Kaveri 37
Goffman, Erving 41–2, 91, 135
Goldberg, David Theo 36
Graeber, David 90
Gragnolati, Michele 130
Gramsci, Antonio 19–22, 167
grief 112–18, 139, 164
grieveability 152, 156–60
Guha, Ramachandra 80
Gupta, Dipankar 24

Hakim, Catherine 153
Hall, Stuart 118
Hammersley & Atkinson 150, 158
Hannam & Diekmann 61, 69, 71
Harvey, David 36, 55, 103
hegemony 19, 106, 167
Hochschild, Arlie 6, 91–2
Hodges, Jessie 9, 30, 33, 76, 122, 124, 130–5, 149
Höijer, Birgitta 105
Holst, Tore 7, 22, 86

informal urbanism 37–40
irony 104, 107, 161–2; of guides 132–5; solidarity 101–2

Jain, P. 65
jhuggi jhopris *see* slum
Johnson, Eleanor 91–2
jugaad 40

Kapoor, Dip 140
Kapoor, Ilan 143
Khair, Tabish 12

Koens, Ko 2, 62
Koven, Seth 17–18, 63, 103
Kvale, Steinar 10

Leys, Ruth 87–8
Lonely Planet 46

Ma, Bob 10, 33
Macaulay, Thomas 63
MacCannell, Dean 42–3, 57
Mayo, Katherine 63–5
Meged, Jane 42
meritocracy 24, 154, 164–7
Meschkank, Julia 18, 34, 39
Miller, Sam 39
Miller, William Ian 71
Mohanty, Sachidananda 61
Morrison, Toni 62
Mukherjee, Meenakshi 13, 56

Naipaul, V.S. 80
Nyers, Peter 69

orphanage tourism 148–9

Pahar Ganj: authenticity 45–6; history 46
pain 112–14, 117, 137–8
Pandey, Gyanendra 20–1
Perkins & Thorns 41
personal story 85, 111, 117, 122–36, 150
pirate modernity *see* Sundaram, Ravi
playful abjection 66–9; on the City Walk 45–6, 69–79, 99
political society *vs.* civil society 22–31, 124
politics of pity/justice 104
Politics of the Governed *see* Chatterjee, Partha
postcolonial studies 21–2, 59–66; in tourism+A10 research 59–61
post-humanitarianism 104–8
Potter, Jonathan 124
poverty porn 101–3, 107–8, 161–2
Pratt, Mary Louise 61
Precariat 151
precariousness 151–4
Public Interest Litigations 65
Puri, Jyoti 65
The Pursuit of Happyness 133

Radjou, Navi 40
Reas, P. Jane 6, 148–9
Richey & Ponte 107

Roberts, Gregory David 39
Rogerson, Christian 2
Roy, Arundhati 61

Said, Edward 12, 43, 142; Orientalism 21–2; tourism 59, 61
Sainath, Palagummi 130
Salaam Baalak Trust staff 75–9, 121–2, 135–7, 141
Sarbin, Theodore 125
Sarda, Shveta 39
Savage, Mike 152
Scheper-Hughes, Nancy 142–3
Selinger & Outterson 33, 67
Sengupta, Mitu 39
sensitisation 111, 116–17, 135–6
shame: of guides 137–8; of society 15, 78; of tourists 112
Sharma, Kalpana 39
shock of the real 66, 144
show/shield debate 33, 55, 137, 161–5
Simpson, Kate 143
Sinha, Mrinalini 63–4
slum 15–18; in Delhi 25–7
Slumdog Millionaire 39, 62, 150
slum tourism: resistance to 59–68, 101–3
Smith, Adam 106–7
solidarity 104, 107, 163–5; of tourists 18, 99, 144, 158
Spivak, Gayatri 18, 20, 22, 124, 143
Standing, Guy 151, 159

Steinbrink, Malte 44, 62; global slumming 17–18, 59
Steinkrüger, Jan-Erik 101–2
subaltern: contemporary subaltern 22–3; honorary subalterns 152, 166–7; liminal subaltern 4–6, 50–1, 74–5, 124–5; Subaltern Studies Group 19–21
sublimity 67, 73, 103
Sundaram, Ravi 26, 168; informal urbanism 37–40, 64–5; pirate modernity 72
sweaty concepts 67–9

tourism performance 41–3, 162
tourist gaze 41–42, 163; and postcolonial studies 59
'touristic borderzone' 43–4; Pahar Ganj 46–7, 61, 66
tourist valorisation 88–93

Urry & Larsen 41–3, 59

Venkatesh, Sudhir 150
Verma, Gita 39–40
Vodopivec, Barbara 142–3
volunteerism 61–2, 142–52, 167

Wetherell, Margaret 123–4
whiteness 62–6
Wright, Carroll 15–18, 137